Crime Mapping Case Studies

Crime Mapping Case Studies: Practice and Research

Editors

Spencer Chainey and Lisa Tompson

University College London
Jill Dando Institute of Crime Science

John Wiley & Sons, Ltd

Telephone (+44) 1243 779777

Email (for orders and customer service enquiries): cs-books@wiley.co.uk
Visit our Home Page on www.wileyeurope.com or www.wiley.com

Other Wiley Editorial Offices

John Wiley & Sons Inc., 111 River Street, Hoboken, NJ 07030, USA

Jossey-Bass, 989 Market Street, San Francisco, CA 94103-1741, USA

Wiley-VCH Verlag GmbH, Boschstr. 12, D-69469 Weinheim, Germany

John Wiley & Sons Australia Ltd, 33 Park Road, Milton, Queensland 4064, Australia

John Wiley & Sons (Asia) Pte Ltd, 2 Clementi Loop #02-01, Jin Xing Distripark, Singapore 129809

John Wiley & Sons Canada Ltd, 6045 Freemont Blvd, Mississauga, Ontario, L5R 4J3

Wiley also publishes its books in a variety of electronic formats. Some content that appears in print may not be available in electronic books.

Library of Congress Cataloging-in-Publication Data

Crime mapping case studies : practice and research / Spencer Chainey and Lisa Tompson, editors.
p. cm.
Includes index.
ISBN 978-0-470-51608-9
1. Crime analysis—Data processing. 2. Geographic information systems. 3. Digital mapping.
I. Chainey, Spencer. II. Tompson, Lisa.

HV7936.C88C747 2008
363.250285—dc22 2007040693

British Library Cataloguing in Publication Data

A catalogue record for this book is available from the British Library

ISBN: 978-0-470-51608-9

Dedications

Spencer Chainey: To Victoria, Oscar and Isaac
Lisa Tompson: To Stephen, Karen and Nik

Contents

List of contributors

Jill Barclay, Technology Manager: GIS, Information Communications and Technology Service Centre, New Zealand Police, PO Box 50040, Porirua, New Zealand

Rachel Boba, Treasure Coast Campus, Florida Atlantic University, 500 NW California Blvd, Port St Lucie, FL 34986, USA

Ian Bullen, Strategic Analytical Partnership Coordinator, Community Safety Team, Greater Manchester Against Crime, 17–19 The Wiend, Wigan WN1 1PF, UK

David Canter, Executive Director, International Centre for Investigative Psychology, School of Psychology, The University of Liverpool, Eleanor Rathbone Building, Bedford Street South, Liverpool L69 7ZA, UK

Tom Casady, Chief of Police, Lincoln Police Department, 575 S. 10th Street, Lincoln, NE 68508, USA

Clare Daniell, Crime Operational Support, National Policing Improvement Agency, Bramshill, Nr Hook, Hampshire RG27 0JW, UK

Gregory Day, Senior GIS Specialist, Geographic Information Management Systems, 461 Innes Road, Durban 4001, Republic of South Africa

Marcus Ferreira, Lieutenant Colonel and Criminal Analyst, Instituto de Segurança Pública, Avenida Presidente Vargas, n° 817–11° e 16° andar, Centro, Rio de Janeiro, Brazil

Andy Gilmour, Business Analyst, Applications Development, Information Communications and Technology Service Centre, New Zealand Police, PO Box 50040, Porirua, New Zealand

Liz Groff, Assistant Professor, Criminal Justice Department, Temple University, 1115 W. Berks Street, Philadelphia, PA 19122, USA

Derek Johnson, Strategic Analyst, Information Management Unit, Dorset Police, Bournemouth Police Station, Madeira Rd, Bournemouth BH1 1QQ, UK

Tim Mashford, State Intelligence Services, 10/412 St Kilda Rd, Melbourne 3004 VIC, Australia

Ana Paula Mendes de Miranda, Governo do Estado do Rio de Janeiro, Instituto de Segurança Pública, Avenida Presidente Vargas, n° 817–11° e 16° andar, Centro, Rio de Janeiro, Brazil

Alice O'Neill, Higher Intelligence Analyst, NIM Development Unit, West Midlands Police HQ, Lloyd House, Colmore Circus, Birmingham B4 6NQ, UK

Dave Ottiwell, Data and Analysis Development Manager, Community Safety Partnership Team, Greater Manchester Against Crime, Divisional and Partnership Support Unit, 4th Floor, Chester House, Boyer Street, Manchester M16 0RE, UK

Chris Overall, Project Manager, Crime Mapping and Analysis Project, Durban Metropolitan Police Service, Box1172, Durban 4001, Republic of South Africa

Jon Poole, Research and Intelligence Manager, Bath and North East Somerset Policy and Partnerships, Bath and North East Somerset Council, Bath Police Station, Manvers Street, Bath BA1 1JN, UK

Steven Rose, CDRP Manager for Rugby Borough Council, Chief Executives Office, Rugby Town Hall, Evreux Way, Rugby CV21 2RR, UK

Kim Rossmo, Research Professor and Director, Center for Geospatial Intelligence and Investigation, Department of Criminal Justice, Hines Academic Center, Room 120, Texas State University, 601 University Drive, San Marcos, Texas 78666-4616, USA

Lorie Velarde, Crime Analyst, Irvine Police Department, 1 Civic Center Plaza, Irvine, CA 92606, USA

Chris Williams, Community Safety Strategic Partnerships Officer, Safer Merton, 3rd Floor Athena House, 86–88 London Road, Surrey SM4 5AZ, UK

Donna Youngs, Managing Director, International Centre for Investigative Psychology, School of Psychology, The University of Liverpool, Eleanor Rathbone Building, Bedford Street South, Liverpool L69 7ZA, UK

Preface

Crime mapping combines a progressive blend of practical criminal justice issues with the application and research field of geographical information systems (GIS) (Chainey and Ratcliffe, 2005). It is a field that has seen increased growth in both the developed and developing world over the past ten years, with police and crime reduction agencies using it to aid intelligence development, criminal investigations, crime prevention, performance improvement, information sharing and crime reduction.

Crime Mapping Case Studies: Practice and Research helps to document developments and applications of crime mapping by providing real work examples, practical solutions, and the presentation of new techniques that can be applied to policing and crime reduction. Several good texts now exist on the subject of 'crime mapping' (e.g. *GIS and Crime Mapping* by Chainey and Ratcliffe, 2005 and *Mapping Crime: Principles and Practice* by Harries, 1999) that explain many of the approaches, methodologies and techniques that have developed or can be applied to crime mapping. We attempt to add to these by illustrating in this book how crime mapping is being applied around the world.

This book intends to build on the two volumes on *Crime Mapping Case Studies: Successes in the Field* (edited by Nancy LaVigne and Julie Wartell) published by the US Police Executive Research Forum. These volumes discontinued after 2000, but played a key role, particularly in the USA, in helping to raise awareness on how crime mapping can be applied. We aim to take forward the spirit of these publications in this book, and if popular into a continual series that will document many other case studies. A difference between this book and its USA predecessor is that it does not just include 'successes in the field'. Instead it also includes examples that demonstrate problems associated with implementing crime mapping and its application, importantly highlighting mistakes and challenges that others can identify with, avoid and learn from.

Each case study in this book either demonstrates a particular application, analytical technique or new theoretical concept, and is written in a style that is accessible and concise. We also hope that the book provides a focus for helping exchange good practice on what works. Indeed one of our motivations for putting this book together was to help to better document many of the excellent presentations that we see at the UK and US Crime Mapping Conferences each year (see www.jdi.ucl.ac.uk and www.ojp.usdoj.gov/nij/maps). We have therefore eagerly sought out what we have considered are some of the good examples of crime mapping that have been presented at these conferences. In doing so, this has helped meet our desire to encourage practitioners to document their work and share it for others to learn. So, several of the contributions you will read in this book have been written by police and crime reduction professionals, including analysts, GIS officers, public safety practitioners and even a Chief of Police! We have also sourced several other case studies from practitioners that we have come across in our travels. In particular, we thought it important to not just illustrate UK and USA examples (where crime mapping is most developed), but gather contributions from Australia, New Zealand, South Africa and Brazil. We hope that together, these case studies offer something to all readers, regardless of their maturity in developing crime mapping and where they are in the world.

The book also includes valuable contributions from established researchers in crime mapping, providing them with an outlet to capture examples or developments of their work, and for you to learn from them in a publication that is more accessible than academic research journals.

The case studies have been grouped into five parts. Part I begins by describing several examples of 'Developing crime mapping'. We begin in New Zealand where Rick McKee and colleagues from New Zealand Police describe how they have been developing GIS and crime mapping tools to assist their police colleagues in tackling crime. Ana Paula Mendes de Miranda and Marcus Ferreira then describe how they have developed a technique to help overcome the challenging task of geocoding crime records in Rio de Janeiro, Brazil. By addressing this it is now helping them to gather further momentum in demonstrating the value of geographical crime analysis for policing improvements in Rio. This is followed by Tim Mashford, capturing how Victoria Police in Australia have met the challenge in implementing crime mapping within a large law enforcement agency. Part I is then nicely rounded off by Tom Casady, Chief of Police of the Lincoln Police Department, Nebraska, USA. Tom eloquently describes how Lincoln have taken crime mapping to the next level by automating many of the standard reports and functions that can plague the analysts' time, freeing them up for more tasks that take better advantage of their analytical expertise. This crime mapping automation includes 'threshold alerts' that help to keep his troops updated on crime patterns.

Part II presents four case studies on 'Geographical investigative analysis'. Kim Rossmo, former Vancouver Police cop and now at Texas State University describes the principles and methods behind geographical profiling. He also has argument over some of the ways in which several researchers have attempted to measure geographical profiling success. He illustrates these points with an example of a burglary series in California. The practical application of geographical profiling is then added to by Claire Daniell from the UK's National Policing Improvement Agency. Claire discusses many of the challenges that are presented to a geographical profiler in their analysis, and uses an example of a series of sexual assaults in the city of Bath to demonstrate these. Chris Overall and Gregory Day from the Durban Metropolitan Police Service, South Africa provide the next case study by describing their use of a probability grid method for helping to explore the spatial patterns of an armed robbery series. The operational use of geographical analysis for supporting police investigations is also then illustrated by Tom Casady who shows how Lincoln Police applied spatial analysis principles to arrest Roosevelt Erving, a bank robber who had previously gone undetected for over four years.

'Neighbourhood analysis' provides the theme of Part III. Alice O'Neill from the UK's West Midlands Police Force describes how they have been applying crime mapping to assist in the strategic allocation of police and crime reduction resources to effectively implement Neighbourhood Policing and their Community Safety Plan. In particular Alice refers to the Strategic Threat and Risk Assessment Index (STRATi) they have developed to help protect the public, promote community stability and reassurance, and reduce victimization. Ian Bullen then provides the first of two case studies from Greater Manchester (UK), describing how they have used and built upon the Vulnerable Localities Index to assist in the identification and strategic analysis of priority neighbourhoods. Dave Ottiwell, also from Greater Manchester Against Crime takes the theme of neighbourhood analysis further by demonstrating how strategic analysis of offenders and their assessment profiles is helping to inform local strategies for reducing re-offending.

Part IV captures three UK examples that illustrate how survey and visual audit data can be incorporated into crime mapping applications, rather than relying only on recorded incident data. Steve Rose from the Birmingham Community Safety Partnership introduces this theme by demonstrating how they have integrated data from their 'Feel the Difference' survey of local residents to help understand those communities where feelings of fear, concern and worry about crime and anti-social behaviour are highest. He also describes how they have included data from an environmental visual audit to help target the Partnership's reassurance work. Chris Williams (London Borough of Merton) then explores how the fear of crime can be explored at the micro- (street) level by targeting and interpreting residential surveys. This is then followed by Jon Poole who describes the Bath and North East Somerset Community Safety and Drugs Partnership's approach to better understanding incidents of crime and anti-social behaviour that are related to the night-time economy. This case study particularly highlights the importance of this type of survey work for helping to better understand the problems that relate to alcohol influenced incidents.

The final Part, Part V, describes some new techniques that are emerging from research. Derek Johnson from the UK's Dorset Police illustrates an operational application of the 'near repeat' phenomenon to help predict patterns of residential burglary. He describes the research on which it is based and how they have placed it into an operational setting that has resulted in measurable additional reductions in crime. Liz Groff from Temple University, Philadelphia, USA then does a great job in describing the application of agent-based modelling to crime mapping. This simulation technique offers considerable potential in helping to inform environmental criminology theory, but probably more importantly test 'what if' type scenarios and observe their results (such as 'if we deployed an extra 20 patrol officers, what impact would this have on crime levels?'). Rachel Boba then discusses how a technique that has principles in crime mapping can be applied to help assess the vulnerability of local targets to terrorism. Here she applies the EVIL DONE criteria developed by Clarke and Newman (2006). David Canter and colleagues from the University of Liverpool (UK) finish Part V, and the book, by describing the Interactive Offender Profiling System, a tool that is being developed to potentially help exploit recent advances in offender profiling and geographical profiling and offer a more real-time exploration of linked crimes, suspect prioritization, criminal networks, target monitoring and future crime activity.

We would like to thank all the authors for their contributions to this book, and hope that you find this collection of their case studies as helpful in raising your awareness and developing your applications of crime mapping. And if you have a case study you would like us to consider for the next volume then do get in touch!

Spencer Chainey
UCL Jill Dando Institute of Crime Science, London, UK

References

Chainey, S. P. and Ratcliffe, J. H. (2005) *GIS and Crime Mapping*. London: Wiley.

Clarke, R. V. and Newman, G. (2006) *Outsmarting the Terrorists*. Portsmouth, NH: Greenwood Publishing Group.

Harries, K. (1999) *Mapping Crime: Principle and Practice*. United States National Institute of Justice. Available online at http://www.ojp.usdoj.gov/nij/maps/pubs.html

Part I Developing crime mapping

1 Developing geographical information systems and crime mapping tools in New Zealand

Andy Gilmour and Jill Barclay

1.1 The starting point

New Zealand (NZ) Police decided as part of a new records management system (RMS) being developed during the mid-1990s, that a key reporting tool for this system would be geographical information system (GIS) based, that would allow users to query police systems to ascertain patterns of crime, by specific areas in requested time frames.

Based on site visits made in North America, NZ Police information and communication technology (ICT) staff commenced analysis of a client-based GIS package to be tailored to local, personal needs. The approach taken was that ICT would develop the tool and then gauge end-user interest of what would be the final product. Nationwide interest was assessed and the front-end GIS application was developed for an initial rollout of 19 analyst sites throughout NZ Police. The product was delivered in early 1999.

This approach resulted in limited frontline interest due to two main factors. First, because there was no previous exposure to GIS packages, some of the analyst staff could not understand the concept of what the application would potentially offer. They questioned the need for it and, to a lesser degree, questioned how usable it would be. Second, the package was to involve a local cost to each site or station receiving it.

Although the NZ Police ICT Service Centre developed and deployed the tool, each site was responsible for the cost of the customised workstation and printer that was

Crime Mapping Case Studies: Practice and Research Edited by Spencer Chainey and Lisa Tompson

deemed a requirement for the application to operate effectively. This was an obstacle for the project. The local cost was significant and, in some cases, even where analysts were eager to obtain the package, their managers were not willing to spend the money for it to occur.

These factors aside, when the application was completed and ready for deployment, the initial target of 19 sites had resulted in an increase to 33. Word-of-mouth and support from others had seen the target group increase by one-third. After a successful local pilot, the desktop package was initially well received by staff, with some sites quickly producing useful and worthwhile outputs. The analysts at each site were able to import raw crime data, specific mapping data for their area and run customised queries to suit local needs.

Street-level base mapping data was supplied to each site from the NZ Police ICT Service Centre. This proved hard to maintain and manage (particularly with software and data version updates) due to end-users being required themselves to perform the local installs of these data on their custom workstations. Over time, there were a number of incompatibility issues with other NZ Police applications, leading to a high maintenance requirement for this GIS tool. The training provided did not completely meet users' needs, which meant some sites were unable to realise the full potential of the application. Eighteen months after distribution of this package, NZ Police made the decision to decommission this application and replace it with a web-based mapping solution.

1.2 Developing a web-based GIS solution for NZ Police

Other than the escalating cost associated with high maintenance, NZ Police had implemented a focus on thin-client-based applications, which were easier to maintain and provided access for more end-users to ICT tools. The requirement for specialist workstations disappeared and was replaced by solutions that any of the 5500 networked computers could use and support 10,500 staff to view the new mapping application.

The focus then became replicating as many of the existing functions that users had grown accustomed to using, but doing it within a thin-client solution. An inaugural browser-based mapping application called MAPS (Map-based Analytical Policing System) was released on the police network in late 2000. Although there were a number of GIS desktop functions that could not be replicated in this thin-client application, a decision was made to proceed with the decommissioning of the GIS tool anyway.

Even though MAPS was predominantly built for and used by Police intelligence analysts to assist in identifying crime patterns and trends, it allowed basic mapping queries to be compiled with a wizard-based formula that could be conducted by all operational police staff. It provided users access to two key sources of police data:

1. The Police Communications and Resource Deployment (CARD) Centre information. Three centres operate nationally in New Zealand. All public calls for service, general and emergency come through these centres. The three centres transfer data to a common read-only database which MAPS accesses. This allows staff to generate maps of calls for assistance as timely as 15 minutes after the police response and attendance.
2. The National Intelligence Application (NIA). NIA is a national records management system that all NZ Police staff use to record key offence and intelligence information. MAPS accesses a repository version of this data and again, within 15 minutes of being recorded within NIA, MAPS is able to create a map of those recorded events.

As with all mapping applications, MAPS relies on geocoded information to be able to display these locations to the analyst. Whereas the CARD system had always geocoded an address for deployment reasons, NIA did not initially enforce structured location data for crime/incident data entry. This required a cultural change across NZ Police to emphasise the importance of meaningful location data for ICT tools to be able to analyse at a future time. The percentage of accurate geocoded data in these early days was in the vicinity of 50%. It was quickly realised that this did not make for a very complete or accurate crime map.

Within a short amount of time a structured crime location data entry process was mandated within NIA. Along with that mandatory requirement came rules on how frontline staff could and could not record these locations. There was initial resistance within some levels of the police to this change. Constant education and reiteration of the importance of meaningful crime data to create meaningful analysis has seen significant business practice change within the past 5 years. Location data is now conservatively estimated to be in the range of 75–80% accurate. The NZ Police continue to work on their quality control for location-based data. Data quality still varies in consistency across the department, with some areas of NZ focusing on the data entry standards more than others.

1.3 Building on the map-based analytical policing system (MAPS)

The past 6 years since the introduction of MAPS have seen three upgrades involving improved functionality, better architecture platforms and better system performance. The focus has been to constantly bridge the functionality gap between a thick-client desktop GIS and the thin-client web browser application that has created MAPS. Improvements in this area have been made but, technically, some things are still not possible.

In 2004, NZ Police invited Spencer Chainey and Dr Jerry Ratcliffe to look at the MAPS application with the focus being on professional guidance on where the tool should be taken in the future. This proved invaluable with constructive guidance on what was required to better meet analysts' needs. This formed the basis of the functionality added to the following two versions of the application. The first version was warmly received by the analyst audience (see Figure 1.1) and NZ Police are, at the time of writing, releasing the second comprehensive version.

Six years on and ICT are still managing the expectations of high-end analyst users who require the use of a full GIS package. Some of these analysts were previous users of the desktop package and some have come into NZ Police having used desktop

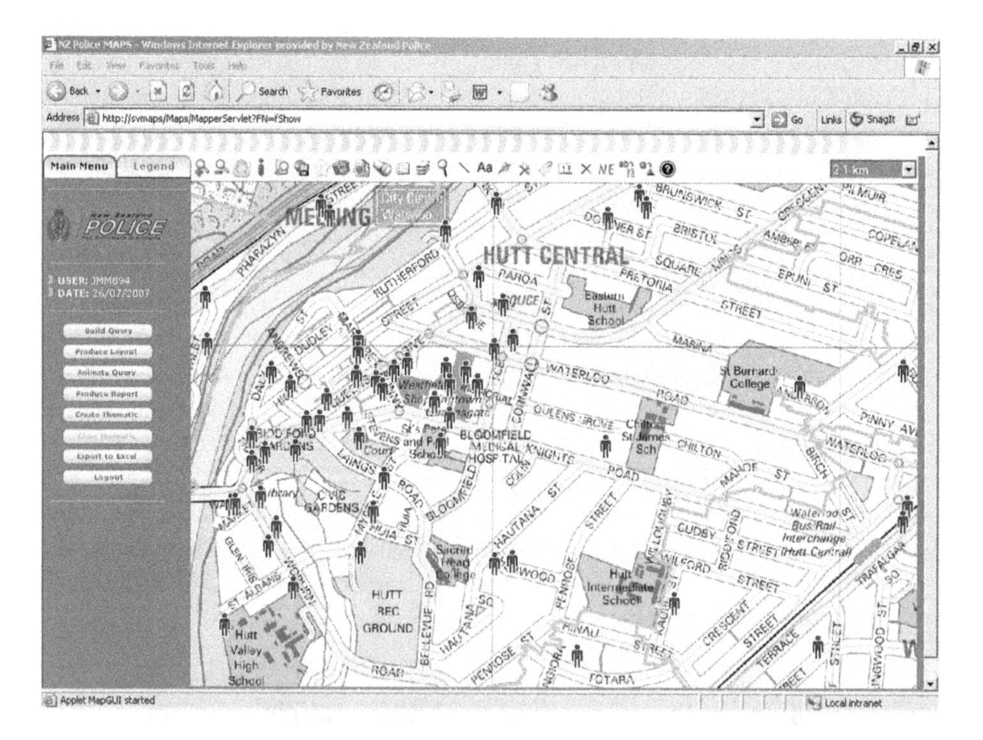

Figure 1.1 NZ Police's Map-based Analytical Policing System (MAPS) (v4.0) interface. MAPS allows users to build a query, select, display and explore crimes for any location in New Zealand. This example shows the Lower Hutt Central Business District with National Intelligence Application (NIA) data for March 2007. The person icons represent recorded NIA 'dishonesty offences where a suspect or offender was nominated'.

GIS in other industry or overseas police departments. This was a point not lost on Chainey and Ratcliffe who had also recommended the need for a reintroduction of a desktop GIS package for a key group of end-users. With enhancements in technology, NZ Police have a strategy to offer the abilities of a desktop GIS to intelligence analysts again. This will utilise the centralised mapping data and provide access to NIA and CARD via the thin client, while providing the GIS analytical tools to the user at the front end.

As part of this future strategy NZ Police will focus on data management, through improved data warehousing structures, and data entry functionality for operational police. The quality of data will continue to be a focus for all of NZ Police with the endeavour of making crime analysis as accurate and useful as possible, while working towards the NZ Police vision of Safer Communities Together.

2 An analytical technique for addressing geographical referencing difficulties and monitoring crimes in Rio de Janeiro, Brazil

Ana Paula Mendes de Miranda and Marcus Ferreira

2.1 Introduction – developments in crime analysis in Rio de Janeiro

Developments in crime analysis in Rio de Janeiro have come about from an endeavour to better structure and organise information originating from two police institutions into a central data facility. In Brazil, unlike many countries, policing is performed by two main agencies: the Military Police who respond to incidents, perform patrolling duties and other activities associated with law enforcement; and the Civil Police who operate as a judicial police agency and investigate crime. The main aim in moving towards a centralised data facility has been to integrate research methods and academic rigour into an institutional assessment of police work through appraisals, target definitions, evaluations, and to create a consistent performance measurement framework that helps contribute to improvements in policing and supports police professional management.

This task has been accomplished by a multidisciplinary technical team of civil and military police officers, and academic researchers with a range of skills, at the Instituto de Segurança Pública (ISP is an independent governmental agency created in 1999 to support public security in the state of Rio de Janeiro. For more details see Miranda, 2006 or visit www.isp.rj.gov.br). The team at ISP produce monthly reports

Crime Mapping Case Studies: Practice and Research Edited by Spencer Chainey and Lisa Tompson

of criminal events and other incidents related to public security for the State of Rio de Janeiro. Data is sourced via the Operational Control System of the Legal Police Station Programme (Programa Delegacia Legal). Forty per cent of additional data are also sourced via special arrangements with the police agencies to help fill any data gaps.

Work on crime analysis started in 1999, as part of the Programme of Statistical Upgrading and Relations with the Media. Many different sectors of the public safety society contributed to its design, including researchers engaged in studying themes such as violence, criminality and public security (Governo do Estado do Rio de Janeiro, 2000). The main goals were to make crime data transparent and accessible, to bring in expertise in statistical data analysis, and to be able to report on changes in crime patterns by Public Security Integrated Areas (AISP). The AISP are geographical areas that cover a Military Police battalion and one or more Civil Police districts. The State of Rio de Janeiro is divided into 40 AISPs.

At present, several statistical reports are produced, showing geographical and temporal crime patterns, and areas that are experiencing trends that are above regional patterns. These reports help to plan police operational actions as well as preventive strategies. They bring together large volumes of information, with an emphasis on using quantitative crime analysis methods to contribute to supporting decision making and a fair distribution of available resources for public security. According to Beato Filho (2000), this use of crime statistics is fairly unique in Brazil. He also highlights that most security secretariats in Brazil lack data collection and statistics facilities. Additionally, a survey carried out by ISP in 2004 identified that only four out of 27 states (Rio de Janeiro, Rio Grande do Sul, São Paulo and Minas Gerais) are regularly updated and informed with crime statistics.

Crime analysis in the State of Rio de Janeiro utilises databanks of offence records collected at the police stations in the State. The records contain information that includes the type of criminal event, time and date, and characteristics of offenders, victims and the location where the offence occurred.

In Rio de Janeiro we experience problems similar (but probably more extreme) to many other places when it comes to geographically referencing crime incidents. These are mainly associated with the difficulty in accurately identifying and recording the location of the event, and record input errors. These problems lead to crime data records lacking sufficient detail to be positioned precisely and accurately to the location where the offence occurred. In addition, problems that relate to the lack of data input standards for street and place names and errors in spelling confound geographical referencing issues.

So, even though there is fairly easy access to computer tools such as a geographical information system (GIS) for supporting the mapping and analysis of crime incidents, the potential use of these tools has been impaired by the absence of crime records with geographical coordinates and the poor quality of recorded crime data. In Brazil the lack of up-to-date and comprehensive cartographic base maps has also caused restrictions in developing geographical crime analysis.

2.2 Analysis by space–time monitoring cells

Despite the poor quality of available information, a geographical analysis tool that enables us to explore certain aspects of localised crime patterns has been developed. The tool is sensitive to the quality of crime data that are available and at least allows us to begin to explore changes in crime patterns. The technique involves the space–time monitoring of geographical cells (CEMET-Cells Monitora Espacio-Temporal) and uses a probabilistic method for geographically referencing each crime incident.

Using a grid matrix of 300 m by 300 m cells that covers the State of Rio de Janeiro, the process makes use of the available address data fields that are held in each crime record and calculates a geographical reference for the crime incident and assigns it to a determined grid cell. The result of this process is a calculated score for each cell based on the probability that the crime occurred in the geographical area covered by the cell. The following examples help to further explain this process.

- If a crime data record contains complete address information such as '55 Rua A', the location of the record can be identified precisely and assigned to its relevant cell with a probability of 1 (see Figure 2.1a)
- If a crime data record contains only partial address information such as 'Rua A', the location of the record cannot be identified precisely. Many conventional geographical referencing processes would leave this record ungeocoded. The CEMET methodology, however, makes use of the available address information to determine, in a probabilistic manner, where the offence may have occurred. 'Rua A' is a long street and covers 11 cells. The CEMET model therefore assigns a probability of 1/11 from this crime record to each cell based on the knowledge that the crime could have occurred with equal probability across any one of the 11 cells crossed by Rua A. This is shown in Figure 2.1b and shows that 'Rua A' extends across cells B8 to L8 and therefore a score of 1/11 is assigned to each cell to indicate the probability of where the crime occurred.
- In a similar manner, if a crime record contains only the name of a district in which the crime occurred, the locational probability value for where the crime occurred that is calculated by the CEMET model is based on the number of cells that cover the area of the district. If, however, both the street name and district are known, both pieces of information are used to determine the location probability score. For example, if the crime data record contains the street name

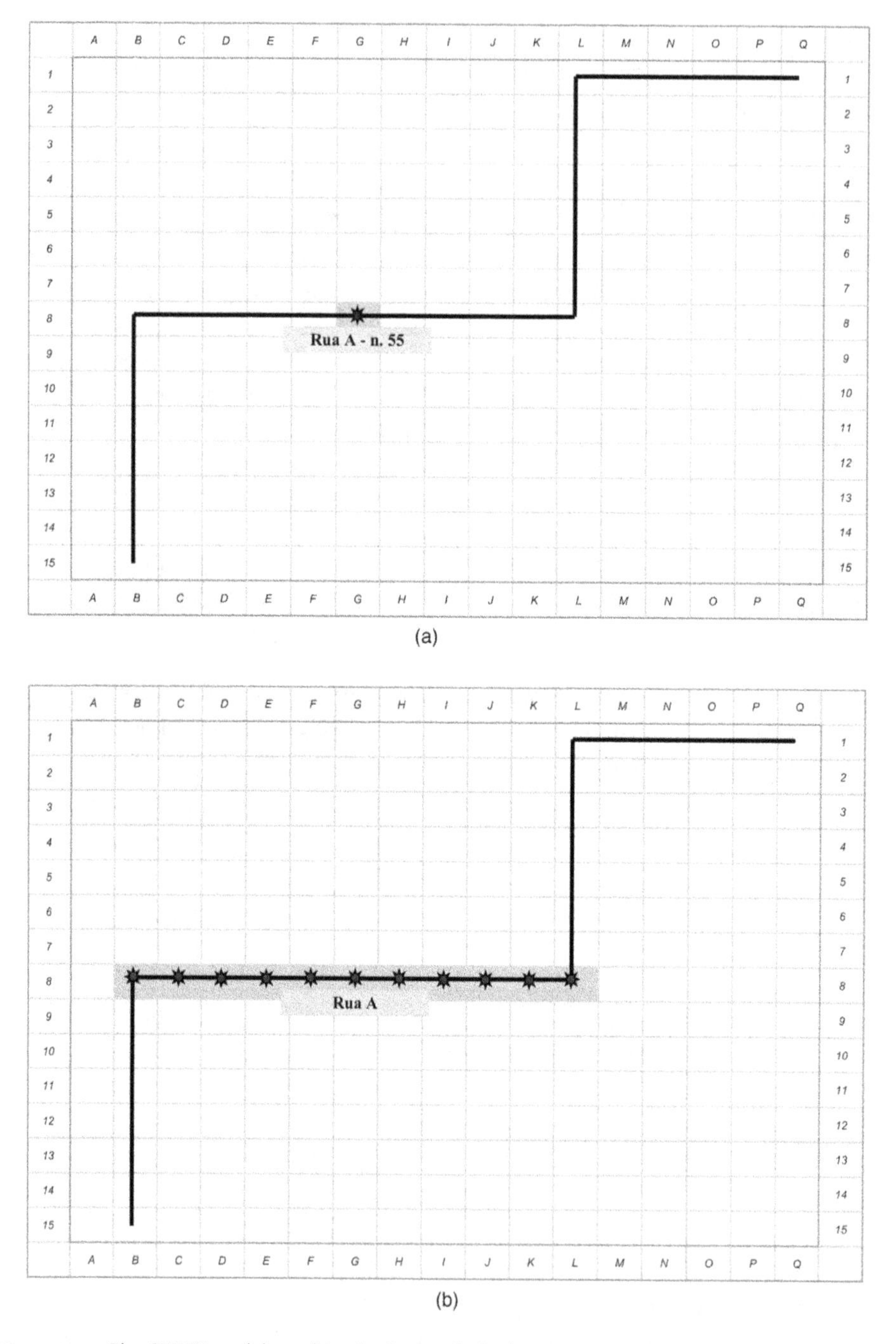

Figure 2.1 The CEMET model used in the State of Rio de Janeiro to determine the geographical positioning of crime incidents. The model uses an approach that calculates a location probability score based on the completeness of address information that is available for determining where

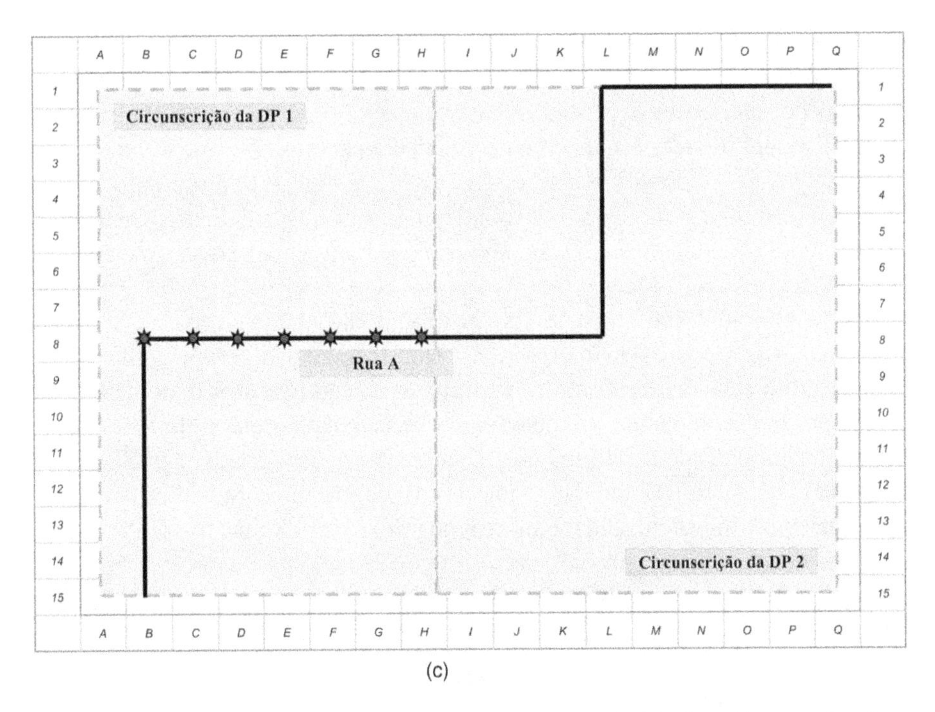

(c)

Figure 2.1 (*Continued*) the crime occurred. (a) A crime record with the address 55 Rua A can be precisely located to cell G8. Cell G8 is therefore assigned a location probability score of 1 for this crime incident. (b) A crime record that contains the address 'Rua A' cannot be pinpointed to a precise location. Rather than leaving this record ungeocoded, the CEMET model determines a probability score of 1/11 based on the number of cells that Rua A covers. Each cell that Rua A covers is assigned this probability score. (c) A crime record with incomplete address information, such as just the street name, may also contain other useful address information that could help to better pinpoint where the crime occurred. Using this address data together (e.g. a street name and a district name), the CEMET model can calculate a probability score that more precisely determines across which cells the crime incident occurred.

and the name of the district, this may help to further pinpoint the location of the incident. This is illustrated in Figure 2.1c. 'Rua A' and district '1' were recorded in the crime record. 'Rua A' also partially extends into district '2'. Using the CEMET model we can determine a location probability score of 1/7 because seven cells cover the section of 'Rua A' that is located in district '1'.

The individual probability scores assigned from each crime record to each cell are then summed to provide a measure of crime incident levels across Rio de Janeiro.

By using the CEMET model, it maximises the crime data that are recorded in Rio de Janeiro (rather than discarding the many crime records that contain incomplete address information and cannot be located precisely), so at least we can begin to explore geographical patterns of crime in a reasonably robust manner. Crime records can also be selected by type and other parameters (e.g. date) and applied to the CEMET model in order for specific geographical patterns of crime to be explored. This enables us to identify where crime levels are high (i.e. those cells with the highest probability score), compare crime trends across Rio de Janeiro, and explore the dynamics of crime by analysing how crime patterns change over time and move between areas.

The definition of incidence levels above regional patterns requires a parameterisation. A procedure was therefore adopted to establish an upper threshold to identify those cells that had scores that were significantly higher than regional levels. Upper threshold values are determined by calculating the arithmetic mean of the regional incidence level plus two standard deviations for the period of crime data that is being considered. By using this approach, those cells that have values above the upper threshold, and therefore high levels of crime, can be identified easily (within a GIS).

The CEMET model has been developed to be independent of digital base maps being available and can be implemented using conventional paper maps provided that these at least show the street network, quarters, districts, zones or any other usable geographical references of interest that can be used to help determine where crime incidents have occurred. This facility offers those areas with limited technology resources and digital base maps (such as in many of the more rural areas of the State of Rio de Janeiro) to be able to explore geographical patterns of crime. Indeed the ISP CEMET model prototype was based on a paper street map of the City of Rio de Janeiro (on sale at any newsstand in the City) and was only later replaced by digital base maps after the model was seen as offering value and when resources became available.

The CEMET model currently used at ISP divides the State of Rio de Janeiro into approximately 500,000 cells of 300 m by 300 m, and is the foundation for generating the following products.

2.3 Identifying crime patterns using paper maps

When GIS software is not available the CEMET model can still be used. In the example shown in Figure 2.2, the crime incident levels were computed by using a software tool developed by the ISP team, with the calculated results shown in a Microsoft Excel spreadsheet against a backdrop of district boundaries. Each cell is shaded according to its probability score to indicate those areas experiencing high levels of crime.

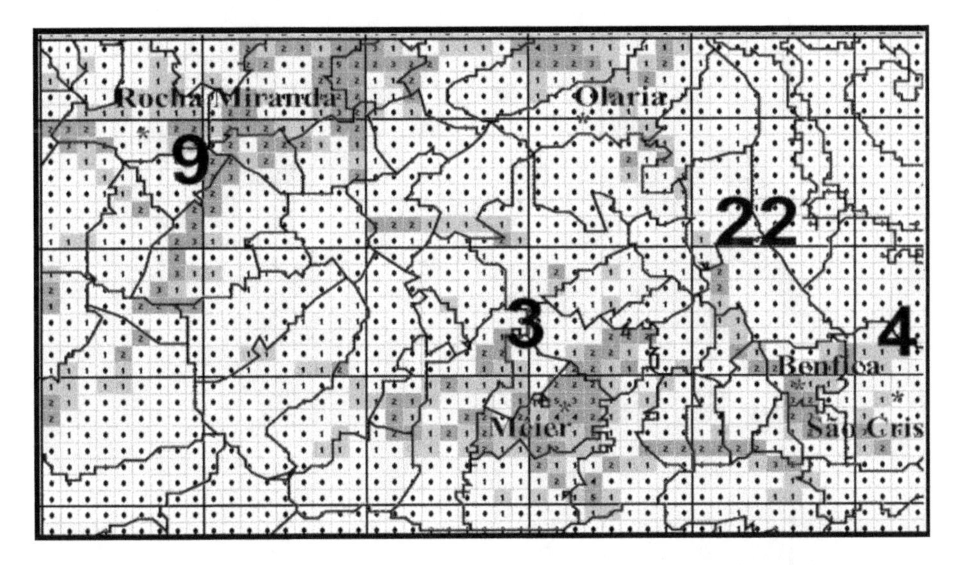

Figure 2.2 The CEMET model can be calculated without the use of a GIS, with results shown in Microsoft Excel. This figure shows crime levels (shown in each cell and reflected in the colour shading) calculated from the CEMET model and illustrated in a Microsoft Excel spreadsheet against a backdrop of imported district boundaries for an area within the State of Rio de Janeiro. A full-colour version of this figure appears in the plate section of this book.

In this case each Microsoft Excel cell represents a 300 m^2 area. An example of the geographical areas that these cells cover is shown in Figure 2.3. This area was identified as a high crime area for street robbery. This area's proximity to a large train station can be noted from the map and aerial photograph. The identification of this site as a high street robbery area was then considered by the police and resulted in the area receiving greater policing patrol attention. An attempt was also made to help better organise the street trading in this area. Continued monitoring at this site has seen a reduction in street robbery after the targeting of this new activity.

2.4 Identifying crime patterns in Rio de Janeiro using GIS and digital cartographic base maps

By using ArcGIS and digital maps of the State of Rio de Janeiro, the CEMET model has been applied across the entire State. Aerial photographs were also added to help interpret the results of the CEMET analysis. As an example, Figure 2.4 illustrates the incidence of recorded offences on a day when a big football game was taking place at the Maracanã football stadium, in the City of Rio de Janeiro. Colour gradations

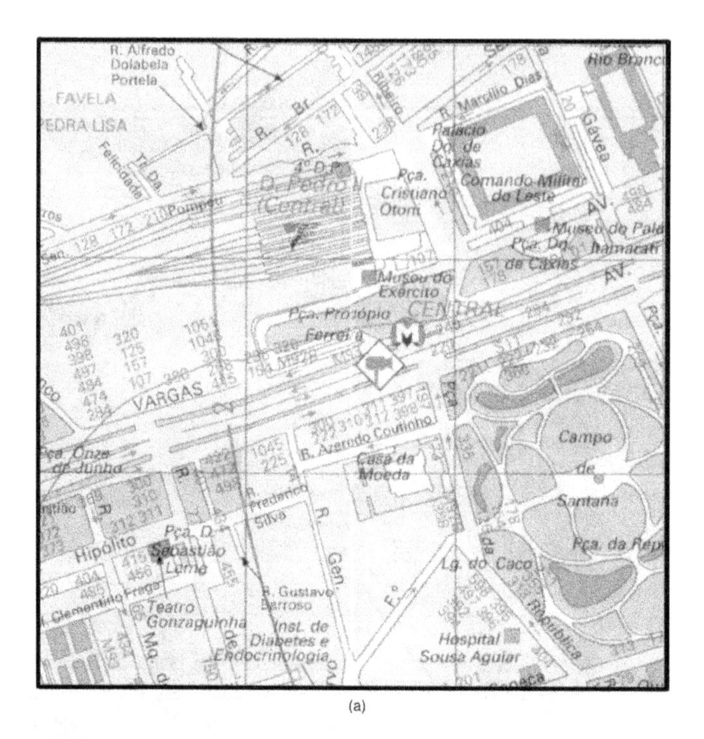

(a)

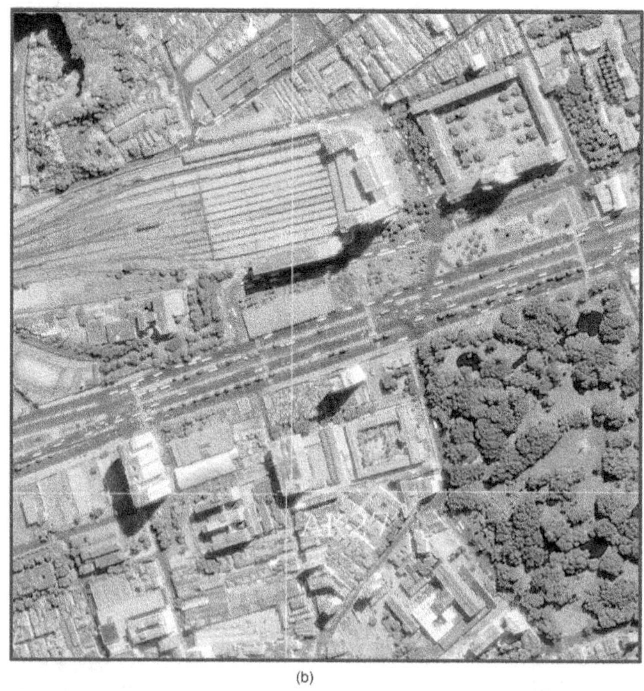

(b)

Figure 2.3 (a) Base map and (b) aerial photograph of an area identified using the CEMET model that has been experiencing high levels of street robbery.

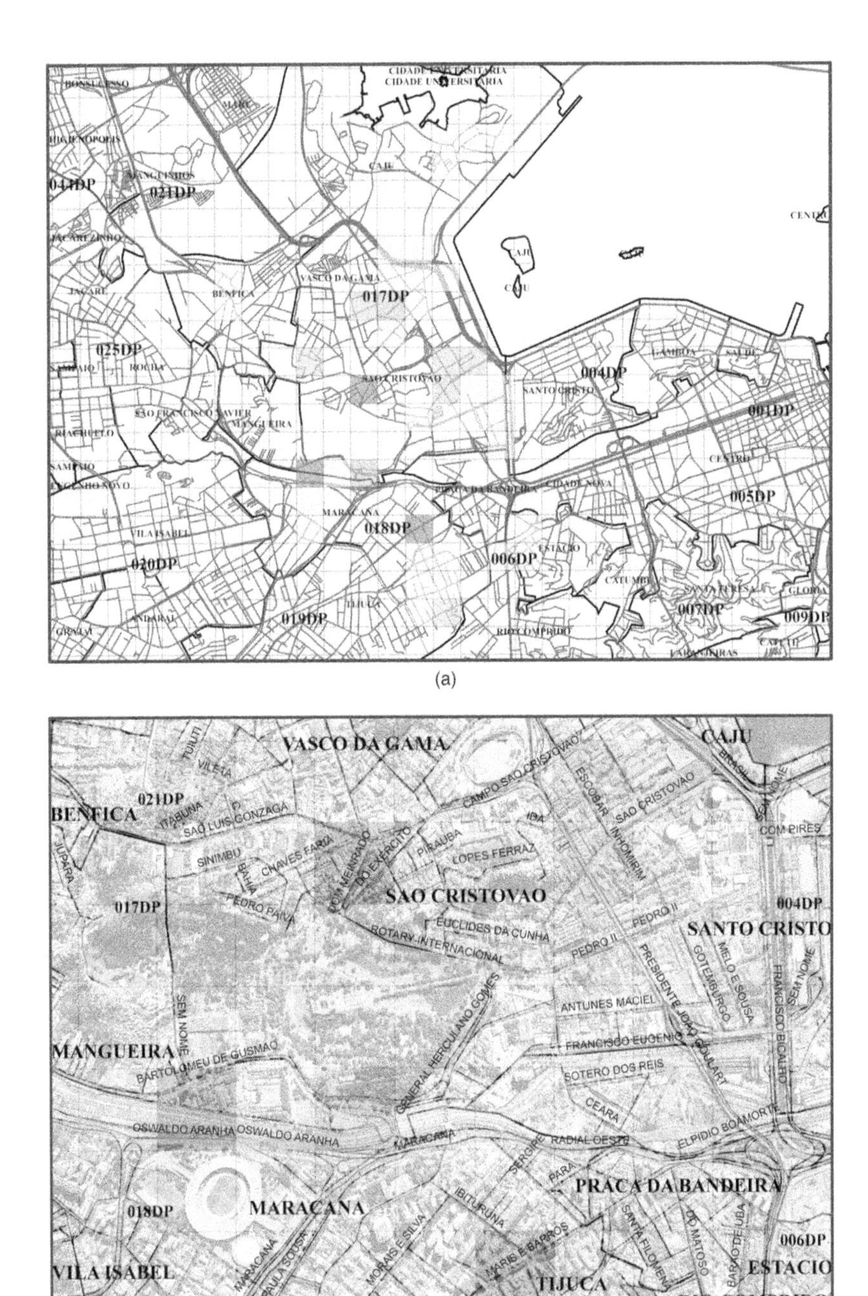

Figure 2.4 Results from a CEMET model output showing crime levels across the City of Rio de Janeiro on (a) a cartographic base map and (b) an aerial photograph on a day when there was an important football game at the Maracanã stadium. Colour gradations indicate the probability score measures for the grid cells. Areas not shaded have not experienced any crime, whereas the darker cells indentify those areas experiencing most crime. A full-colour version of this figure appears in the plate section of this book.

indicating probability score measures can be illustrated on a cartographic base map (Figure 2.4a) and an aerial photograph (Figure 2.4b). With this type of analysis it is possible to support security plans for future events, as well as assessing the effectiveness of policing in the area during, before and after these events.

2.5 Crime analyses on bus routes in Rio de Janeiro

This analysis made use of grid cells to investigate which bus routes are subjected to disproportionately higher levels of crime. Crime incidents were first processed so that probabilistic scores of each cell could be calculated. Following this, a cross-check against bus routes was performed, assigning to each bus route the sum of the cell scores along which each bus runs. Thus, the result of the calculation was a probability score for each line from which a report could be generated. This method has already been used to identify the most vulnerable bus routes and guide the deployment of specific preventive police activity.

2.6 Conclusions

Analysis by space–time monitoring cells (CEMET) was developed due to the difficulties in geographically referencing crime records in the State of Rio de Janeiro. We believe that its use in supporting crime analysis can contribute to a better understanding of crime patterns in order to support better policing (through better targeting patrols and improved investigations) and for helping managers and policy-makers to improve forecasts and plan against future scenarios. This approach is providing the State of Rio de Janeiro with a useful basis on which to develop crime analysis and helping the State to develop improvements in its public policies and improved police professionalism.

2.7 References

Beato Filho, C. (2000) Fontes de Dados Policiais em Estudos Criminológicos: Limites e Potenciais. *Segurança Pública no Brasil: uma discussão sobre bases de dados e questões metodológicas.* Rio de Janeiro: UCAM/IPEA.

Governo do Estado do Rio de Janeiro (2000) *Política Pública para a Segurança. Justiça e Cidadania: plano estadual.* Rio de Janeiro.

Miranda, A. (2006) Informação, análise criminal e sentimento de (in) segurança: considerações para a construção de políticas públicas. Soares, Andréia (org.) *A análise criminal e o planejamento policial.* Rio de Janeiro: Instituto de Segurança Pública, Vol. 1.

3 Methods for implementing crime mapping within a large law enforcement agency: experiences from Victoria, Australia

Timothy Mashford

3.1 Introduction

As the primary law enforcement agency for the state of Victoria, Australia, Victoria Police covers an area that is roughly the size of Great Britain. With over 13,600 personnel, Victoria Police ranks as one of the largest police forces in the world, servicing a population of over 5 million.

Geographical information systems (GIS) have been in use within Victoria Police since the mid-1990s, primarily in support of intelligence capabilities. In particular, the State Intelligence Division has utilised GIS software since 1994 in order to map and analyse crime locations. In 1997, GIS capabilities were enhanced with the introduction of the Geographic Intelligence Unit (GIU), which was established after a member of the State Intelligence Division travelled abroad and viewed how various other law enforcement agencies were successfully using GIS.

One of the first projects undertaken by the GIU was the development of a custom tool to simplify the use of MapInfo (Troy, NY) GIS software. Written using MapBasic, and borrowing from interstate examples, the Geographic Analysis System (GAS) included simple push-button functions to facilitate using the software, as well as some basic spatial analysis tools. Its ease of use promoted GIS within Victoria Police, and by the year 2000 there were seven full-time staff dedicated to GIS

Crime Mapping Case Studies: Practice and Research Edited by Spencer Chainey and Lisa Tompson

in two separate units (Intelligence and Information Technology – IT). There were also over 200 MapInfo licences throughout the force, primarily within the Divisional Intelligence Units (DIU). However, over the next few years the use of GIS diminished significantly, primarily as a result of departmental restructuring and loss of critical staff. By late 2003, there was only *one* dedicated GIS staff member.

In mid-2004, the GIU acquired three experienced personnel, who were tasked with improving the use of GIS throughout Victoria Police. By this stage there were over 230 licences of MapInfo software, however, very few were in use due to lack of user training. The role of the unit was redefined so as to focus on spatial analysis, and as such was renamed the Geospatial Analysis Unit (GAU). In particular, the GAU was being requested to assist with major investigations through the application of Geographical Profiling techniques, after it achieved successful outcomes with several cases. However, before the unit was able to take up this role, it had to ensure the divisional units had the capability to perform their own GIS function.

This task of implementing crime mapping at so many locations across the state required a plan.

3.2 A phased plan for development and delivery

One of the key goals in implementing crime mapping across Victoria Police was to achieve consistency across the State. The plan decided upon was a phased approach in order to help strengthen the development and use of standards, and the ways in which GIS was used to support police practice.

The first phase was the development of a set of standards for crime mapping in Victoria Police. This provided a framework for how GIS was to be used, and would steer the process to ensure consistency. Some of the standards were already in use, and simply needed to be defined. Other standards, however, were decided upon based on sound research and consultation. Similar documents concerning GIS standards were also obtained in consultation with, and from, international agencies, in order to ensure best practices were used.

The standards defined included:

- *Data* – key data sets were defined as the base map layers. Although there was flexibility as to which layers were to be used, the base map layers provided a consistency throughout the force.
- *Coordinates* – the VICGRID94 coordinate system was chosen for storing digital spatial data, as it covers the entire State (Victoria is covered by multiple UTM (Universal Transverse Mercator) zones for spatial coordinates).
- *Temporal analysis* – the Victoria Police database (LEAP) records incidents with a 'From' and 'To' date/time. To avoid skewed results when using either of these

values (or the 'Mid' time), the aoristic method (Ratcliffe, 2000) was chosen to be used for temporal analysis.

- *Geocoding* – crime data were to be geocoded using specific reference files, including address points and ranges. This would help to ensure that geocoding was accurate to within 200 m.
- *Hotspots* – the kernel density estimation method was used to define a hotspot (clustering) of point data.
- *Symbology* – a set of symbols were defined for representing crime categories on a map. This provided consistency between offices.
- *Layouts* – printed maps were to include key cartographic elements including title, scale bar, north arrow, legend and security rating.

The next phase was consultation with GIS users, those being mainly DIU members. The GAU visited several DIU offices to observe how GIS was being utilised, and to discuss their key requirements. This provided an excellent opportunity to understand where the crime mapping process was getting stuck, and in several cases it was found that most crime records were being geocoded manually – a time consuming task which prevented analysis from occurring. After collating this information, the GAU organised a user group meeting to present the Crime Mapping Standards document, and discuss its impact on GIS users.

Once the requirements had been determined, the GAU began development on a new version of the original GAS custom tool. Known as the Geographic Analysis System version 3 (GAS3), the tool was written in-house using MapBasic. As well as the custom program, GAS3 included data sets (vector base map layers and raster imagery) in a defined directory structure to be placed on the workstation. Because the Victoria Police network covers such a large geographical area, it was not possible to implement a central spatial data server.

The functionality of GAS3 includes:

- *Geocoding* – the GAS3 geocoding routine runs through several processes in order to best locate the record.
- *Base map* – standard base map with preset styles, labelling and zoom layering.
- *Imagery* – as the user clicks on the map, this tool determines the correct imagery tile and adds this to the map window.
- *Locator* – allows the user to locate specific places using commonly used formats (address, intersection, street directory grid reference).
- *Selection* – several tools simplify the structured query language (SQL) dialogue, including a 'keyword' finder and table joiner.

- *Spatial analysis* – functions include centroid, proximity analysis, stolen–recovered vehicle linker, and repeat location finder.
- *Layouts* – predefined layouts at varying page sizes, including common elements such as the Victoria Police badge, title styles and a standard disclaimer.

As well as GAS3, a commercial program, HS_Gridder (Portolan Technology) was obtained to provide hotspot and temporal analysis capability. HS_Gridder uses kernel density estimation to determine point data clustering, and creates a shaded surface (for an example see Figure 3.1a). HS_Gridder also applies aoristic analysis techniques to create temporal charts, allowing peak times to be easily determined (see Figure 3.1b).

After trialling GAS3 at several sites with positive feedback, GAS3 was deployed to all GIS users by Victoria Police's IT provider. This required consideration for the best way to distribute the data, as the base map layers and raster imagery required a large amount of disc space (over 5 GB). DVD discs were distributed, and included a small command file, which created the directory structure and copied the data onto the hard disc drive. This common directory structure ensured data updates could be easily disseminated, until a time when a central data server could be made available.

When GAS3 was being deployed around the state, the GAU set about developing a training programme for Victoria Police analysts. Key functions from MapInfo, GAS3 and HS_Gridder were identified that would allow an intelligence practitioner to map and analyse crime data at a basic level. The format of the training programme was determined through consultation with analysts who had previously received training in crime mapping. An earlier outsourced 'pilot' training course had been provided over two sessions – two days of basic GIS training, followed by a month to practice, and then two days of more advanced training. However, feedback from the participants suggested that many did not get the opportunity to use GIS during the practice period in between, and arrived at the second session having forgotten everything from the first! Although this format was therefore deemed unsuitable, it was felt that providing four consecutive days of training was too intense for novice users.

The GAU decided that the training would run for two days, during which the basics of crime mapping and analysis were presented and practised. Since participants would be presented with a substantial amount of information during this brief period, and many would not put this into practice immediately (as had occurred during the pilot course), it was therefore necessary to provide a comprehensive training manual that could be used as a reference when back at their office. A standard training manual was obtained from MapInfo Australia, however, feedback from the previous courses had indicated that members required training content to be contextualised,

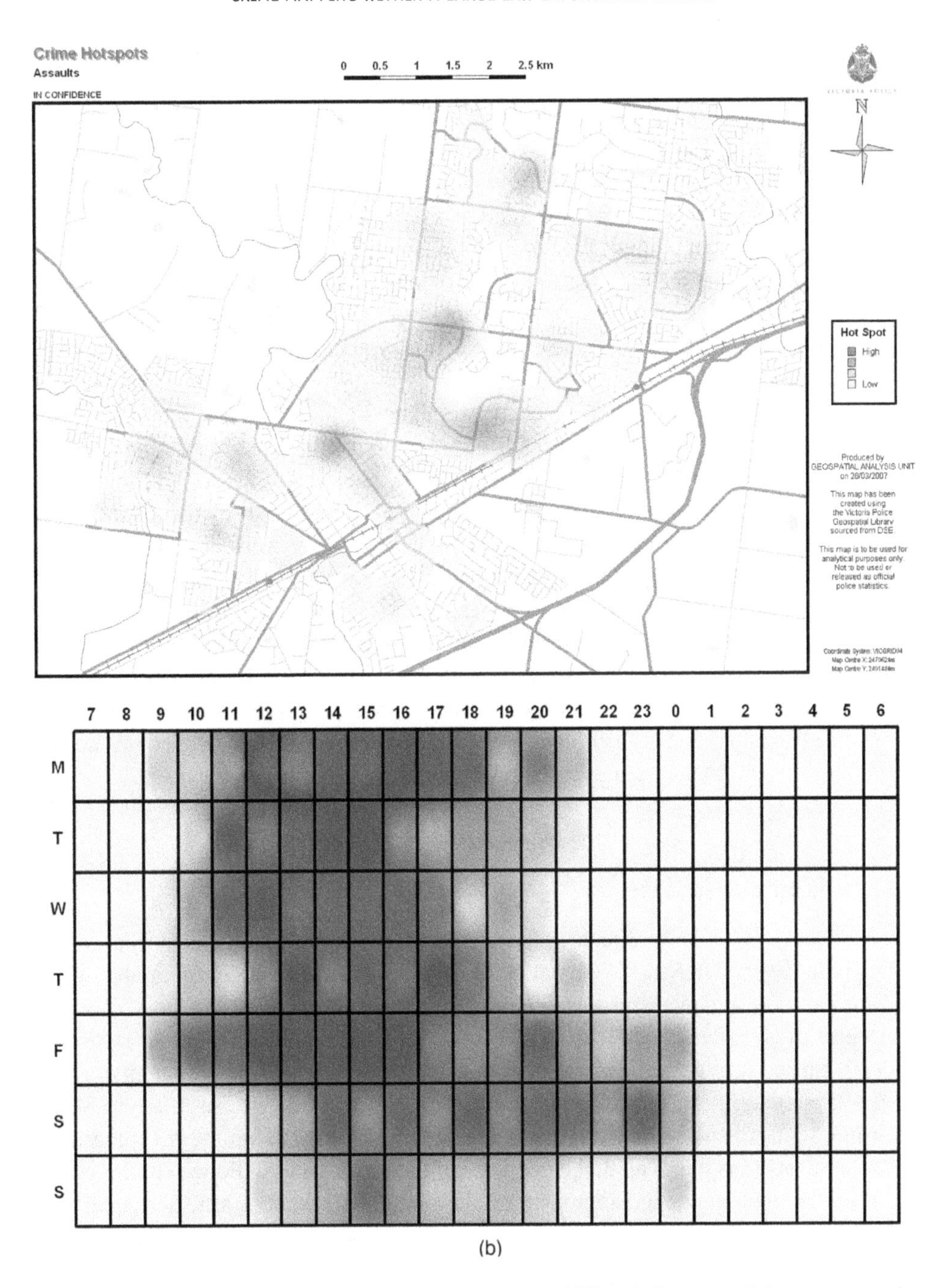

(b)

Figure 3.1 (a) Kernel density estimation hotspot maps and (b) aoristic temporal charts generated using the HS_Gridder software tool used by Victoria Police. A full-colour version of this figure appears in the plate section of this book.

i.e. using police scenarios and data. The training manual was therefore altered to use data specific to Victoria Police, and also included GAS3 and HS_Gridder functions.

Numerous practical tasks were included in the training programme, in order to reinforce the techniques and concepts being presented. At the end of the two-day course, each participant was instructed to complete an assessable task in which they were presented with a crime problem scenario that required the application of crime mapping and analysis techniques in order to achieve an outcome. As well as producing an analytical product, participants were also required to record the steps that were taken to reach their outcomes. Members of GAU then assessed each final product.

The first set of training courses occurred in late 2005, during which time over 80 members attended the course. Participants were a mix of police officers and civilian staff, and there was a large variance in experience levels and IT skills. The course syllabus was constantly revised during this period, based on participant feedback. Several topics, such as writing SQL Select queries, were removed as they took up substantial amounts of time and caused confusion for delegates.

In order to encourage attendance by all areas, no cost was imposed onto the participants; however, members from country areas had to provide their own transport and accommodation. The GAU also travelled to several country areas in order to provide the training on site, which was done in order to ensure country areas were not disadvantaged due to their locations. Other areas received customised (reduced) training specific to their duties, as appropriate.

3.3 Progress to date

As of early 2007, over 200 members have received the training in basic crime mapping. Every DIU and Regional equivalent are now utilising crime mapping in some capacity, and many on a daily basis. There are now over 300 MapInfo software licences in use, all installed with GAS3 and HS_Gridder. The type of crime mapping being carried out still varies, however, this is due to individual requirements rather than limited abilities. For example, tactical intelligence officers based in the State Crime Squads usually limit their crime mapping activity to simple mapping and aerial imagery, as this is all that is required of them. As Victoria Police continues its shift to an intelligence-led approach, smaller units such as the Traffic Management Units and the Criminal Investigation Units are being resourced with a dedicated intelligence analyst, who as part of their role performs crime mapping.

As the Divisional Intelligence Units have become comfortable with the basics of crime mapping, the GAU has developed an advanced training programme that has recently begun to be rolled out. This programme focuses on analysis techniques, including an introduction to environmental criminology. Furthermore, updates to base map data and the GAS3 program are still provided via distributed discs, however, the common directory structure ensures that these updates can still be completed by the member without requiring complex actions. Presently, a study is underway regarding the feasibility of a central spatial data server on the internal network.

The increased use of crime mapping has made it a common process in Victoria Police operations, and demand for its use has grown significantly. Currently, a working group is examining web-based mapping solutions, which could be provided over the intranet for general use by all members.

3.4 Crime mapping projects – some examples

Since the divisional offices are now mostly self-sufficient with regards to crime mapping, the GAU has been free to focus on its analytical role with the State Intelligence Division. Several of the major projects completed recently include:

- *Mercury 05* – the 2005 Multi-Jurisdictional Exercise (MJEX) involved law enforcement agencies, the Australian Defence Force, and intelligence agencies across Australia. The GAU contributed to the State Intelligence Division's response capability through mobile transmission of spatial data, for intelligence collection purposes. Field operatives transmitted information using a PDA device, and these data appeared on a web mapping system for viewing within the main operations centre.
- *Melbourne 2006 Commonwealth Games* – this was the largest security operation ever undertaken by Victoria Police. The GAU, as part of the Commonwealth Games Intelligence Office, provided a capability for incident response.
- *Vehicle Theft Reduction Strategy* – the GAU analysed Theft Of/From Motor Car incidents in order to identify key areas to target in this strategy (see Figure 3.2). Also as part of this ongoing project, the GAU provides district hotspot maps, which are placed onto the Victoria Crime Stoppers website for public viewing.

3.5 Conclusions

Despite the fact that the implementation of crime mapping within Victoria Police was a time consuming task, it has nonetheless been a highly successful venture. With such a large police force, spread over a considerable geographical area, the coordination

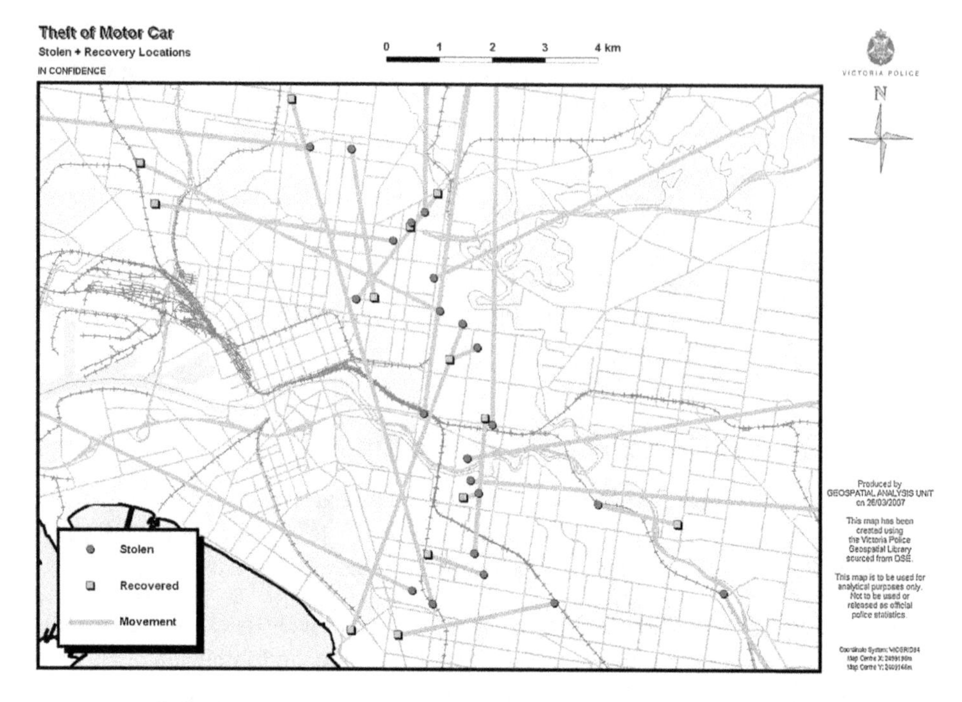

Figure 3.2 Theft of motor vehicle analysis included exploring the relationships between where the vehicle was stolen and where it was recovered.

of crime mapping practices has proven to be a challenging task. However, a well planned approach has ensured that consistency has been achieved across the state. Essentially, it has established crime mapping as a vital tool in support of intelligence analysis and police operations within Victoria Police.

3.6 Reference

Ratcliffe, J. H. (2000) Aoristic analysis: the spatial interpretation of unspecific temporal events. *International Journal of Geographical Information Science* **14**(7): 669–679.

4 Automating briefings for police officers

Tom Casady

4.1 Introduction

Much of what crime analysts and crime mappers do for police briefing purposes can be automated to varying extents. Automating the production and distribution of crime maps, displays, reports and other analytical products can result in more timely and more valuable information to the users, freeing crime analysts from repetitive tasks so that their time and skills can more often be applied to complex and intensive work.

In our daily lives, we regularly encounter information feeds that are automatically updated from computer databases. The electronic sign on the platform lets us know when the next northbound train will be arriving. The monitor at the airport terminal informs us that our departing flight is delayed. The television channel runs a banner across the bottom of the screen providing sports scores, weather forecasts and the current performance of the stock market. Applying some of this same approach to crime analysis is a valuable next step for police agencies that have developed mapping and analysis capabilities.

Automatically updating police information is not new. In many police stations, the information technology staff have deployed tabular database reports that can be set to run and print without human intervention. Thus, in police agencies around the world the commander arrives at work to find the overnight log of incidents in his or her printer tray. In this rather rudimentary example, a program has been written that queries the data to find all the dispatch records within the past 24 hours. The program instructs this command to run at 06:00 hours, and directs the output to the printer. These semi-automated printed reports are common, and there may be many more sophisticated automated routines around other police stations. As the tasks

Crime Mapping Case Studies: Practice and Research Edited by Spencer Chainey and Lisa Tompson

are the same every time, developing a program to automate the process is fairly straightforward.

The same is true of some (but not all) mapping and analysis tasks: the work to be done is essentially composed of the same steps, repeated at a regular interval. For instance, an analyst may produce a weekly printed map depicting changes in crime within police district boundaries: the districts with increases of more than 10% are shaded in red, those with decreases of more than 10% in blue, and the remainder in light grey. The steps necessary to repeat this each Monday morning are the same. With a little planning and thought, this task can be automated, so when the analyst arrives at work they find the freshly produced map on their laser printer – ready for distribution.

As quite a number of crime mapping and analysis products are regular reports, the prospect of automating these has great potential. Think for a moment about the repetitive nature of the queries. We usually want to know the same thing we wanted to know last year, or last week, or yesterday, or six hours ago: Where are the burglaries concentrated? Which felons on conditional release are residing in District 3? How do violent crimes this month compare with violent crimes during the preceding month? A huge amount of the work of analysts in police agencies is directed at repeatedly querying the same dataset in the same way to produce the same map, chart and table and the same report – just for a different time period.

4.2 Automating crime mapping outputs in Lincoln Police Department

The Lincoln Police Department is a municipal police force in Lincoln, Nebraska, USA. Lincoln is a city of 242,000 with a police force of 317 officers and 105 civilian employees. Lincoln has automated many mapping tasks to leverage the capabilities of its police database and its geographical information system (GIS) in several ways. Lincoln's automation of geographical briefing reports is primarily based on the concepts of 'thresholds' and 'alerts'. When a given level (the 'threshold') of activity occurs in a pre-defined geographical area, the appropriate personnel receive an email (the 'alert') advising them of the activity, which includes both a map and a tabular report with the details (see Figure 4.1). Other than the original set up, no human intervention is required. Commanders, supervisors, detectives and even patrol officers can determine the parameters of the alert. Here are several examples:

4.2.1 Alerting officers to increases in crime

Thefts from automobiles are a significant crime problem in Lincoln, and tend to be clustered in both time and space. Every day at 05:00 hours, the police GIS queries

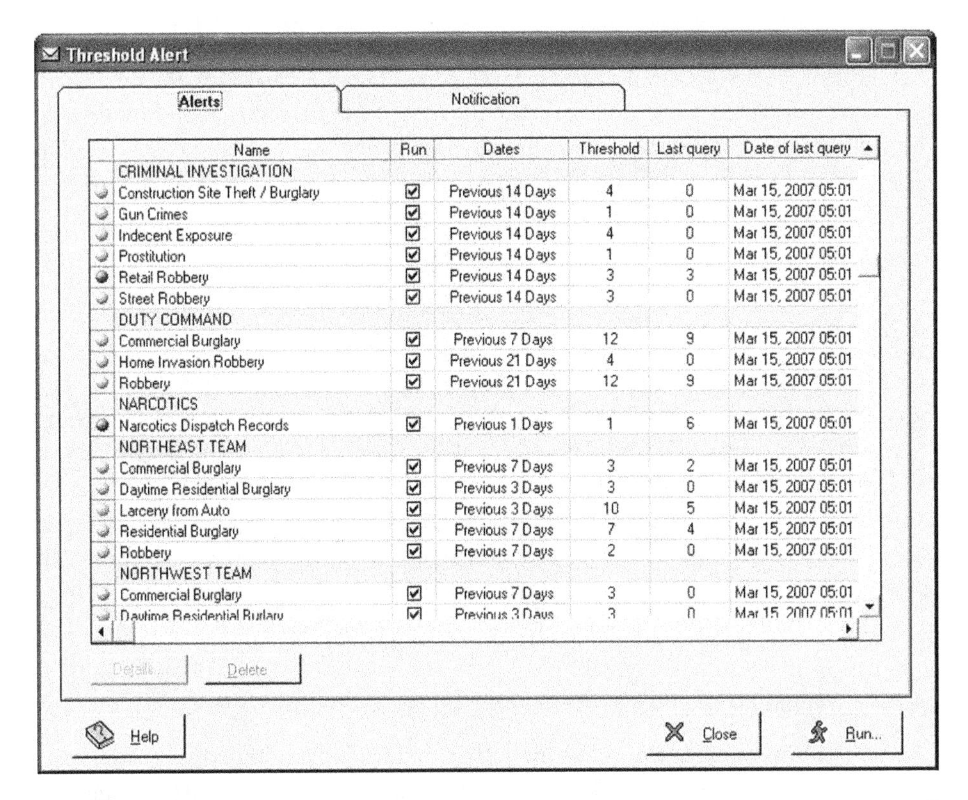

Threshold Alert

Alerts | Notification

Name	Run	Dates	Threshold	Last query	Date of last query
CRIMINAL INVESTIGATION					
Construction Site Theft / Burglary	☑	Previous 14 Days	4	0	Mar 15, 2007 05:01
Gun Crimes	☑	Previous 14 Days	1	0	Mar 15, 2007 05:01
Indecent Exposure	☑	Previous 14 Days	4	0	Mar 15, 2007 05:01
Prostitution	☑	Previous 14 Days	1	0	Mar 15, 2007 05:01
Retail Robbery	☑	Previous 14 Days	3	3	Mar 15, 2007 05:01
Street Robbery	☑	Previous 14 Days	3	0	Mar 15, 2007 05:01
DUTY COMMAND					
Commercial Burglary	☑	Previous 7 Days	12	9	Mar 15, 2007 05:01
Home Invasion Robbery	☑	Previous 21 Days	4	0	Mar 15, 2007 05:01
Robbery	☑	Previous 21 Days	12	9	Mar 15, 2007 05:01
NARCOTICS					
Narcotics Dispatch Records	☑	Previous 1 Days	1	6	Mar 15, 2007 05:01
NORTHEAST TEAM					
Commercial Burglary	☑	Previous 7 Days	3	2	Mar 15, 2007 05:01
Daytime Residential Burglary	☑	Previous 3 Days	3	0	Mar 15, 2007 05:01
Larceny from Auto	☑	Previous 3 Days	10	5	Mar 15, 2007 05:01
Residential Burglary	☑	Previous 7 Days	7	4	Mar 15, 2007 05:01
Robbery	☑	Previous 7 Days	2	0	Mar 15, 2007 05:01
NORTHWEST TEAM					
Commercial Burglary	☑	Previous 7 Days	3	0	Mar 15, 2007 05:01
Daytime Residential Burlary	☑	Previous 3 Days	3	0	Mar 15, 2007 05:01

Details... | Delete

Help | Close | Run...

Figure 4.1 The tabular Threshold Alerts that are automatically generated for Lincoln police staff to show crime levels and if any have risen above the specified threshold.

the incident report database for thefts from automobiles in the Northwest Team Area (one of the police department's geographical command areas). If more than ten of these crimes have occurred in the previous seven days in the Northwest Team Area, the commander receives an email with two attachments. The first is a map depicting the location of the offences, created with the department's GIS software, ArcGIS (ESRI, Redlands, CA). The second is a tabular database report containing the pertinent details: time, date, location of occurrence, the investigating officer, case status and the case number. The report is created with Crystal Reports, and sent in Adobe (Adobe Systems, Inc., San Jose, CA) portable document format (.pdf).

4.2.2 Following up investigations

A Detective Sergeant (DS) coordinates the follow-up investigation of residential burglaries. Her workweek begins on Sunday at 06:30 hours. On Sunday mornings at 06:15 hours, the police GIS queries the incident report database for all residential burglaries reported in the preceding seven days. The DS can quickly examine the map for obvious clusters, and can obtain the basic facts from the tabular report. The case number is a hyperlink – clicking on it sends her into the full police case file. The map and report are not on the printer; rather, the officer will find them in their email inbox, ready for review.

4.2.3 Monitoring sex offenders

By law, the City of Lincoln prohibits high-risk registered sex offenders from residing within 500 feet of a school. Each morning at 04:00 hours, the police department's GIS software queries the sex offender database for registered sex offenders whose address has been updated within the preceding 24 hours to see if the new address is within 500 ft of a school. If any records match these parameters, a map and report are generated and emailed to the crime analysis unit. No one needs to constantly check for violators.

4.2.4 Neighbourhood policing: keeping an eye on prolific offenders

An Officer's assigned police beat includes a small neighbourhood with some significant crime problems. He likes to keep the heat on petty criminals, and he uses the police department's arrest warrant files to do so. The police department holds thousands of arrest warrants for minor offences, such as failing to appear in court in response to a traffic citation, littering or possession of marijuana. Jeff (one of our neighbourhood cops) believes that petty offenders often are also involved in more serious crimes – thefts, burglaries, drugs and so forth. If he can find time to conduct some 'knock and talks' he can occasionally arrest a person wanted on a warrant. Even if he cannot find the person, the mere fact that he was around looking for him or her may cause the offender to lay low for a few hours or a few days – a good result in itself. The addresses of offenders on arrest warrants are notoriously outdated, but the most recently issued warrants are the best prospects. When Jeff arrives for his work week at 14:30 hours on Tuesdays, he has a map and report waiting in his email inbox of all the people in the neighbourhood that have new arrest warrants issued within the past week. He can print a copy for his clipboard, or save it on his memory stick for his in-car computer. Moreover, if he's reviewing the electronic document, he need only click on the name to launch a web browser and be taken to the police department's master name index with all the details about that person and his or her prior police contacts.

4.3 Developing the automation of tasks in Lincoln

These examples are part of the forty-seven daily or weekly 'threshold alerts' that are automated at the Lincoln Police Department, covering everything from prostitution arrests to gun crimes. Forty-seven is a moving number though, as the process of creating, deleting and modifying threshold alerts is dynamic. Essentially, officers who are interested in being automatically notified of certain time-and-space events, or conditions, can work with the department's crime analysis unit to set up a threshold alert that serves their purpose. After the alert is set up, it runs itself on the schedule that was established. It runs when the crime analysis unit is closed, when the employee is on vacation, in the small hours of the morning and during holidays.

The department's 'threshold alerts' are created using CrimeView, a software suite from the Omega Group (San Diego, CA). Special software, however, is not necessarily needed to start automating mapping and analysis tasks. Quite a bit can be accomplished with simple freeware and shareware programs. Many PC applications include tools such as macro recorders that allow users to either partially or fully automate tasks that must be completed on a regular basis. The software is not the important part; it is the logic. If a query is frequently run, a map regularly produced, a particular website frequently accessed, there is probably a way to automate this so it can happen quicker, earlier and without the presence of the analyst.

Although it can certainly help free some time, the value of automating some of the department's crime mapping work is not primarily making life more leisurely for the GIS staff. Rather, automation can help disseminate the information more quickly and reliably.

Consider this example: each Monday and Friday, a crime analyst produces a map of armed robberies. A short narrative description is written of the most recent trend information, and a bulletin containing the map, text and chart is created from a desktop publishing template. The resulting document is printed and distributed to shift supervisors and area commanders via intradepartmental mail. Rather than printing and distributing the bulletin, the process could have been automated – or even just partially automated: if you can send a document to a network printer, you could send the document to a designated directory on a web server. Virtually all desktop publishing and word processing software supports export as .html and/or .pdf formats that are ready to post for Internet/intranet access. Then, everyone has access to the document immediately – and from anywhere they can connect to the Internet and reach the URL. You could email the link to the recipients, or merely update the content to which a link on an internal web page points.

In the previous example, only the distribution of the map, not its production, was automated. This alone, however, has eliminated the delay caused by intradepartmental mail delivery. It has provided access to an unlimited number of people, it has provided an electronic version of the bulletin that can be retrieved at any time, and it has avoided the need to make second-generation photocopies for further distribution. No one needs to return to their office to see the bulletin; it cannot be lost in the mail or under the pile on the desk. You could access it from a networked PC at headquarters, with an air card in a laptop, or even with a web-enabled cell phone or smartphone.

4.4 Automating crime mapping in your agency

Automating crime mapping is not an all-or-nothing proposition. Start small, and more opportunities will become evident as time passes. A good starting point would be to look in your existing software packages for automation tools. In ArcGIS, for example, learn about the macros, saving and loading queries, and the powerful Model Builder. Try out some of the freeware and shareware products for recording, saving and replaying macros: basically, about any function you can perform with keystrokes and mouse clicks can be created as a macro or script which can be replayed at will. Check out Windows task scheduler (Scheduled Tasks in the Control Panel), a handy utility for setting up the unattended running of executable files. Finally, think about how you can automate the distribution of mapping products by pushing these to consumers with email, hyperlinks and web services.

Getting actionable information into the hands of the police personnel actually in a position to act upon it is the key to great crime analysis. A rough drawing on the back of an envelope showing a night shift street sergeant just where in the downtown area the ATM robberies at bar break are occurring is far more valuable than the prize-winning GIS map layout produced a week later with the same information in glorious colour. Automation can help deliver basic mapping products quickly to those who can act upon the analysis.

Part II Geographical investigative analysis

5 Geographic profiling analysis: principles, methods and applications

D. Kim Rossmo and Lorie Velarde

5.1 Introduction

Geographic profiling is an investigative methodology that uses the locations of a connected series of crimes to determine the most probable area of offender residence (Rossmo, 2000). Its main function is to prioritise suspects and assist in investigative information management. Originally applied in cases of serial murder, rape and bombing, it is now increasingly used in serial robbery, burglary, arson and fraud investigations. As a consequence, methods, training and computer software have been developed specifically to deal with the application of geographic profiling to property crime – a methodology now generally referred to as geographic profiling analysis (GPA). This chapter outlines the underlying theories, assumptions, considerations, software and investigative strategies involved with GPA. A case study is presented and evaluation results discussed.

5.2 The theoretical principles behind geographic profiling

Geographic profiling was developed from research conducted at Simon Fraser University's School of Criminology in the early 1990s. Environmental criminology, in particular Crime Pattern Theory, Routine Activity Theory and Rational Choice Theory (Brantingham and Brantingham, 1981, 1984, 1993; Cornish and Clarke, 1986; Clarke and Felson, 1993; Felson, 2002, 2006), provide the conceptual basis for geographic profiling. Crime locations are not distributed in space randomly, but

Crime Mapping Case Studies: Practice and Research Edited by Spencer Chainey and Lisa Tompson

rather are influenced by the features and road networks of the physical environment. An understanding of these patterns provides a means for determining the most probable area of offender residence.

There is a strong relationship between an offender's search base and the location of their crime sites. For a crime to occur there must be an intersection in both time and place between offender and victim. Crime is therefore the product of the criminal, the victim/target and the physical environment ('where and when'). People, criminals included, have their regular routine activities, such as commuting to work, shopping, and visiting friends and family. These locations, and the travel routes between them, make up a person's activity space or comfort zone. Criminals typically commit crimes in those areas where their activity space overlaps suitable targets. Because criminal offenders possess at least a limited rationality, they consider the rewards, risks and effort involved – including distance travelled – in their target selection (see Rengert and Wasilchick, 2000).

Consequently, the most important influence on where criminals offend is where they go during their non-criminal activities (Bennett and Wright, 1984). Most (but not all) crimes occur less than two miles from an offender's residence (Rossmo, 2000). On the other hand, predatory criminals are less likely to commit their crimes too close to home because of a desire for anonymity. A mathematical representation of this understanding was encoded in the Criminal Geographic Targeting (CGT) computer algorithm, now used in geographic profiling (Rossmo, 1995). The CGT model produces probability surfaces that outline the most probable area of offender residence. These are displayed through colour geoprofile maps that provide a focus for investigative efforts.

5.3 Geographic profiling methodology

Geographic profiling cannot solve crimes – only physical evidence, a witness or a confession can do that. Its role is to assist investigators in more effectively and efficiently reaching one of these resolutions. Geographic profiling should be regarded as one of several tools available to detectives, and is best used in conjunction with other investigative methods.

5.3.1 Assumptions and considerations

A geographic profile is based on certain assumptions which, if violated, might affect the accuracy of its results.

- The linkage analysis for the crime series is accurate and reasonably complete (i.e. the same offender committed the linked crimes, and there are not a significant number of unlinked crimes that should be part of the series).

- The offender is a local hunter, not a poacher (i.e. the offender is not commuting into the area to commit the crimes).
- If there is more than one offender, they reside together or in the same area.
- The offender's search base has not changed during the time period of the crime series (i.e. the offender has not moved).

Although it is important to have knowledge of the above assumptions and why they are important, their violation does not always preclude the use of geographic profiling as compensation techniques are sometimes possible (Velarde, 2005).

A geoprofile may produce two peak areas, an indication the offender has more than one search base. Information regarding land use, zoning and area characteristics helps interpret such outcomes. Alternative anchor points can include work sites, drug dealers, fences, past residences and accomplices.

5.3.2 Rigel Analyst

Rigel Analyst is geographic profiling software based on the original Rigel Profiler system developed by Environmental Criminology Research Inc. (ECRI) (www.ecricanada.com). It incorporates an analytic engine using the patented CGT algorithm, geographic information system (GIS) capability, database management and powerful visualisation tools. Crime locations provide the input and are entered by the optional means of street address, latitude and longitude, or digitisation. This reflects the realities of policing in which crimes can happen anywhere – houses, parking lots, back alleys, parks, mountain ravines, and so on. Latitude and longitude coordinates can be determined from a handheld global positioning system (GPS) that reads the user's position from a satellite fix.

Scenarios, wherein crime locations are weighted based upon certain theoretical and methodological principles, are next created and examined. The software produces a geoprofile, a colour map showing the most likely area of offender residence. Suspect addresses can be evaluated according to their position on the geoprofile (expressed as a hit score percentage on a z-score histogram), allowing the prioritisation of suspects, tips, registered sex offenders and other information. Output can also be viewed in Google Earth (Mountain View, CA), assisting the user in displaying land use and physical structures within the region of interest.

5.3.3 Investigative strategies

Address information is an element of most record systems and a geographic profile can be used as the basis for several investigative strategies (Rossmo, 2006). Although specific approaches are best determined by the police investigators familiar with the case, some examples of tactics successfully used in the past are presented below.

- *Suspect prioritisation.* Suspects with known addresses can be prioritised for follow-up investigative work. The problem in many serial crime investigations is one of too many suspects rather than one of too few. Profiling can help manage large numbers of suspects, leads and tips.
- *Directed patrol and surveillance.* The peak area of the geoprofile can be used to help direct saturation patrol and surveillance efforts. This strategy is most effective if the offender is operating during specific time periods.
- *Neighbourhood canvassing.* Neighbourhood canvassing, information sign posting, community cooperation, media campaigns and area searches, can be focused with a geoprofile. Specific areas can be targeted for leaflet distribution or directed mailings of letters from the police department requesting suspect information from the public.
- *Police record systems.* A geographic profile can be used to search computerised police record systems with address information (e.g. computer-aided dispatch, records management, automated jail booking systems, sex offender registries, etc.). Offender description, *modus operandi*, behavioural profile details and similar information can help modify the search criteria.
- *Other data sources.* Databases are often geographically based, and parole and probation offices, mental health clinics, schools and other agencies located in the peak geoprofile area may provide useful information. Several commercial companies offer law enforcement agencies the ability to search multiple personal information databases.
- *Department of motor vehicle (DMV) searches.* If suspect vehicle information is available, a geographic profile can be used to search registered vehicle and driver's license files from state or provincial DMV records. A multiparameter search (i.e. vehicle type, make, colour and likely owner address) can help narrow down thousands of records to a manageable volume.

5.4 Applying geographic profiling to 'volume' crime: the Irvine Chair burglaries

For several years the City of Irvine in Orange County, California, suffered from a series of residential burglaries in several middle and upper-class neighbourhoods. It was believed a single offender was responsible, but various strategies developed by the Irvine Police Department (IPD) to apprehend him were unsuccessful and the burglaries continued. The IPD's Special Investigations Unit (SIU) finally decided to use an intelligence-led policing approach, combining geographic profiling – then a new resource within the police department – with crime forecasting and directed surveillance.

The known series involved 42 residential burglaries that had occurred over the previous two years (see Figure 5.1). The offender exhibited a consistent pattern of behaviour. He only chose single-family homes, preferably those that backed up to green belts or parks. This allowed him to case houses while still on public property. He established escape options before making entry to the residence. Point of exit was always at the rear. If the back fence was high, the offender placed a chair next to it. Motion lights were unscrewed. Property stolen was generally limited to cash and jewellery, items that could easily be carried in the offender's pockets or a small bag. Glove marks were noted at some of the crime scenes. It was apparent the offender was a professional burglar.

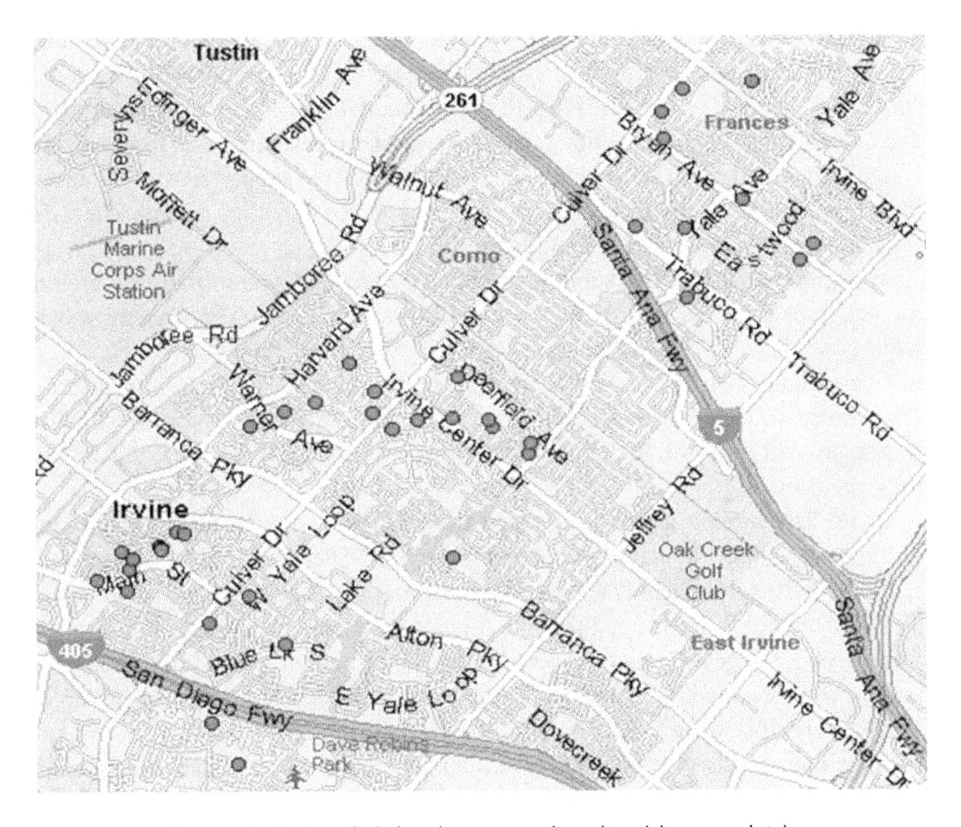

Figure 5.1 Irvine Chair burglary case: crime sites (shown as dots).

Crime Scene Investigators (CSI) had conducted blind swabbing for trace DNA at 21 of the burglaries. The swabs were submitted to the Orange County Sheriff's Department (OCSD) Crime Laboratory for processing. Although it was likely that some of the swabs contained DNA from home owners, it was hoped a common profile – the offender's – would be found at two or more burglaries. In August 2005, the OCSD Crime Laboratory was able to extract DNA from five of the 21 swabs. Four of the five swabs with DNA were profiled as males. Of the four male DNA profiles, three were unique to one person.

Rather than attempt to locate the offender in the process of a committing his next burglary, an act that may only take a few minutes, the SIU decided to look for him as he hunted for his next target, a process that takes much longer. Crime forecasting predicted the offender would most likely commit his next burglary on a Friday, Saturday or Sunday, between 17:00 and 22:00 hours. The SIU then had to determine the best location in the city to deploy its limited surveillance resources; as the crimes covered approximately 17 square miles, a geographical focus was necessary.

A geographic profile was prepared which identified a peak area 0.25 square miles in size. The anchor point for most criminal's searches is their home. But for some, called poachers, the search base is a different location. Irvine is a wealthy community, and most of the professional criminals who commit crimes there come from outside. Neighbourhood demographics in the peak geoprofile area supported the hypothesis that the offender was not local. The geoprofile appeared to outline an area where the burglar began his hunt after commuting to Irvine.

In September 2005, SIU surveillance teams were deployed in the peak area of the geographic profile during the days and times forecasted. The strategy was to identify vehicles by license plate number as they drove through the area. These numbers could then to be used to identify non-local individuals from California Department of Motor Vehicles (DMV) records and traffic citations for investigative follow up. Although the SIU committed to the strategy for five weeks, this proved to be unnecessary.

On the first evening, a vehicle was seen driving in and then out of a neighbourhood in the peak area of the geoprofile (see Figure 5.2). The license plate number was recorded and a subsequent DMV record check revealed the vehicle was a car rental. The SIU members determined that the driver was Raymond Lopez, an ex-convict living in Los Angeles County. According to the rental company, Lopez had been renting vehicles weekly for the past 20 years. Lopez became a surveillance target, and over the following weeks he was observed casing homes in various residential neighbourhoods in Irvine and other Orange County cities. Lopez was tracked through the use of a global positioning satellite (GPS) device, which placed him near residential burglaries in the cities of Cypress and Rossmoor.

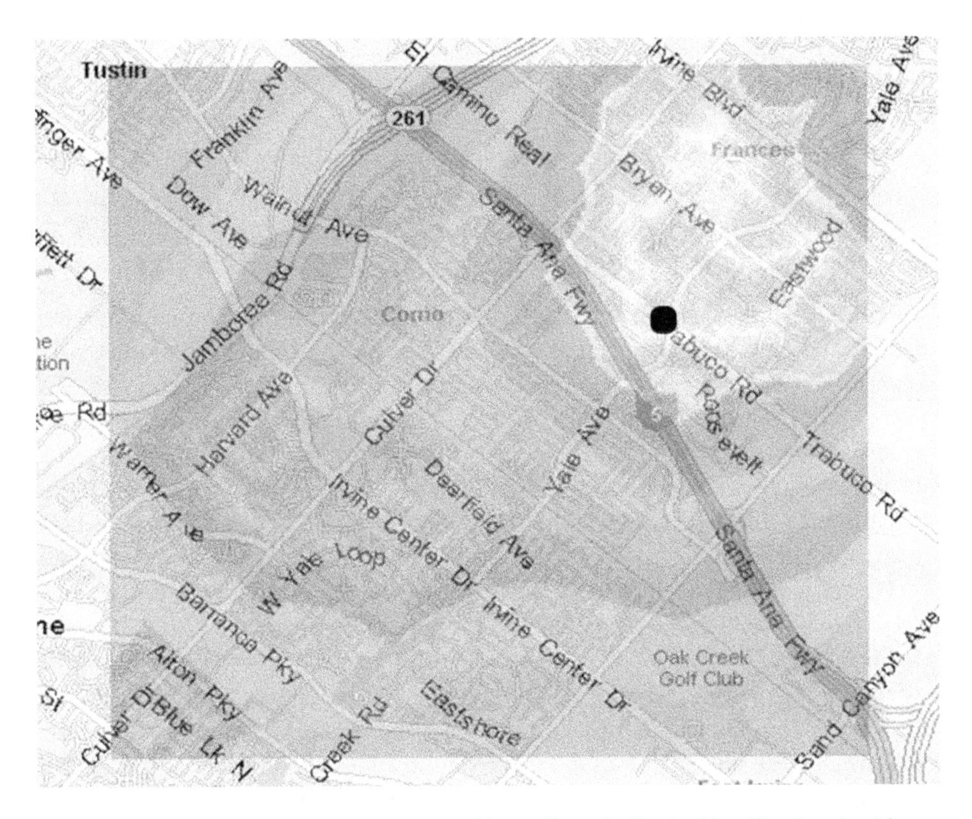

Figure 5.2 Irvine Chair burglary case: geographic profile and offender identification site (shown by the large dot). A full-colour version of this figure appears in the plate section of this book.

Crime Scene Investigation officers swabbed the steering wheel of one of Lopez's returned rental cars and the samples were submitted for analysis. In late October 2005, the Crime Laboratory reported his DNA was a match. Detectives obtained search and arrest warrants. On the evening of 2 November 2005, Lopez was arrested by IPD detectives. He was in possession of property from a burglary he had just committed.

A search warrant was served at Lopez's home and a handgun, cash and other items, including gold bricks made from melted stolen jewellery, were seized. A search warrant was also executed at a pawn shop/jewellery store believed to be acting as a fence for Lopez. Detectives seized over $500,000 worth of rare coins and jewellery. The pawnshop owner told detectives that associates of Lopez had pawned property there regularly for the past 20 years.

Lopez was responsible for 139 burglaries committed between January 2003 and September 2005 alone, with a total property loss exceeding $2.5 million. Other law enforcement agencies have come forward with information suggesting he was responsible for burglaries in their jurisdictions as well. Lopez was charged with 14 felony counts, including burglary, possession of stolen property, and felon in possession of a handgun. In October 2006 he pleaded guilty to 14 felony counts and began serving a 13-year state prison sentence.

5.5 Measuring the effects of geographic profiling in Irvine

The Irvine Police Department has now won three awards for the Chair Burglary case, including the prestigious International Association of Chief of Police (IACP)/ ChoicePoint Award for Excellence in Criminal Investigations. Although not the first agency to implement GPA, it has been one of the more active. During the 19 months following implementation, the IPD Crime Analysis Unit completed 27 geographic profiles, the majority for robbery series. By mid-year 2006, 11 (41%) of these cases had been solved, permitting the accuracy of the geoprofile to be measured (Velarde and Cooper, 2006).

The accuracy of geographic profiling is evaluated using a search-cost measurement known as hit score percentage (HS%). The hit score percentage is calculated by dividing the area that needed to be searched following the geoprofile to find the offender's residence, by the total area covered by the crimes (Rossmo, 1993; in press). The 11 IPD cases in their evaluation had a median of 22 crimes. The median and mean HS% were 2.5% and 5.2%, respectively. By comparison, the median and mean hit score percentages for a spatial mean search strategy were 4.7% and 10.4%, respectively. When the relative search efficiency of the two methods are directly compared, Rigel Analyst identified the offender's anchor point in 32% of the area needed to be searched following the spatial mean strategy. Consistency is also important in an analytical method. The HS% standard deviation for Rigel Analyst was significantly less variable than that for the spatial mean (0.08% versus 0.12%).

In an operational context, geographic profiling performed well at identifying the offender's search base for unsolved cases. Rigel Analyst was more effective than the spatial mean by a factor of over three at locating the offender's anchor point. The performance results are consistent with those from other studies on geographic profiling accuracy (Rossmo, 2000, 2005). Geographic profiling analysis has been used on robbery, burglary, automobile theft, vandalism and stolen credit card cases, and IPD's Crime Analysis Unit is now finding it difficult to meet the demand for GPA service requests.

5.6 References

Bennett, T. and Wright, R. T. (1984) *Burglars on Burglary: Prevention and the Offender*. Aldershot, Hants: Gower.

Brantingham, P. J. and Brantingham, P. L. (1984) *Patterns in Crime*. New York: Macmillan.
Brantingham, P. L. and Brantingham, P. J. (1981) Notes on the geometry of crime. In *Environmental Criminology*, Brantingham, P. J. and Brantingham, P. L. (Eds), pp. 27–54. Beverly Hills: Sage.
Brantingham, P. L. and Brantingham, P. J. (1993) Environment, routine and situation: toward a pattern theory of crime. In *Routine Activity and Rational Choice*, Clarke, R. V. and Felson, M. (Eds), pp. 259–294. New Brunswick, NJ: Transaction.
Clarke, R. V. and Felson, M. (Eds) (1993) *Routine Activity and Rational Choice*. New Brunswick, NJ: Transaction.
Cornish, D. B. and Clarke, R. V. (Eds) (1986) *The Reasoning Criminal: Rational Choice Perspectives on Offending*. New York: Springer-Verlag.
Felson, M. (2002) *Crime and Everyday Life*, 3rd edn. Thousand Oaks, CA: Sage.
Felson, M. (2006) *Crime and Nature*. Thousand Oaks, CA: Sage.
Rengert, G. F. and Wasilchick, J. (2000) *Suburban Burglary: a Tale of Two Suburbs*, 2nd edn. Springfield, IL: Charles C. Thomas.
Rossmo, D. K. (1993) Multivariate spatial profiles as a tool in crime investigation. In *Workshop on Crime Analysis through Computer Mapping; Proceedings*, Block C. R. and Dabdoub, M. (Eds), pp. 89–126. Chicago, IL: Criminal Justice Information Authority.
Rossmo, D. K. (1995). Place, space, and police investigations: Hunting serial violent criminals. In *Crime and Place: Crime Prevention Studies,* Vol. 4, Eck, J. E. and Weisburd, D. L. (Eds), pp. 217–235. Monsey, NY: Criminal Justice Press.
Rossmo, D. K. (2000). *Geographic Profiling*. Boca Raton, FL: CRC Press.
Rossmo, D. K. (2005) Geographic heuristics or shortcuts to failure? Response to Snook et al. *Applied Cognitive Psychology* **19**: 651–654.
Rossmo, D. K. (2006) Geographic profiling in cold case investigations. In *Cold Case Homicides: Practical Investigative Techniques*, Walton, R. (Ed.), pp. 537–560. Boca Raton, FL: CRC Press.
Rossmo, D. K. (in press) Geographic profiling evaluation methodology. *Police Practice and Research: An International Journal.*
Velarde, L. (2005) Accuracy and value of geographic profiling in an operational context. Paper presented at the *Crime Mapping Research Conference*, Savannah, GA, September.
Velarde, L. and Cooper, J. (2006) *Geographic Profiling Analysis: Implementation Assessment and Review*. Unpublished manuscript, Irvine Police Department: Irvine, CA.

6 Geographic profiling in an operational setting: the challenges and practical considerations, with reference to a series of sexual assaults in Bath, England

Clare Daniell

6.1 Introduction

Geographic profiling is an investigative support technique designed to provide assistance in cases of serial violent crime (Rossmo, 2000). Geographic profilers attempt to determine the most probable location in which an offender resides or works. Various systems have been developed to give this prediction (e.g. Rigel, Dragnet, CrimeStat). Operationally, in the UK, a system called Rigel is used. This utilises the criminal geographic targeting (CGT) model, which has been developed into a computerised geographic profiling system. Profilers can provide investigators with a probability surface that highlights the likelihood of the offender having an anchor point at any location within a statistically defined hunting area in which the crimes have occurred (Rossmo, 2000). Its conception is well founded in the theories and research of environmental criminology, and the algorithm ensures that the process has a degree of objectiveness. Rigel is not 'an "x" marks the spot' routine, but rather offers an optimal search strategy – with the resultant 'peak' area of the profile indicating the area where the offender is most likely to have an anchor point.

Crime Mapping Case Studies: Practice and Research Edited by Spencer Chainey and Lisa Tompson

This case study demonstrates some of the challenges faced by geographic profilers during the production of an operational geographic profile. In this chapter the issues that need to be overcome in relation to linkage and accurate scene identification are discussed and the application of geographic profiling in relation to offenders who commute into an area to commit offences is examined with reference to a series of indecent assaults in Bath, England. Analysis of the temporal and spatial data of the offences, combined with the geographic profile, supported inferences made by investigators regarding the offender's hunting activity. The case study illustrates that statistical geographic profiling technology should be used in combination with other more qualitative techniques and analysis in order to effectively assist investigations. That is, geographic profiling is not just a tool that predicts the location of an offender's residence in isolation of more qualitative spatial analysis – a misconception by many researchers attempting to evaluate the success of geographic profiling. Additional analysis is then required to interpret the profile in order that it can be applied operationally. This analysis entails an in-depth consideration of influencing factors, including the demographics of each offence location, transport networks, the time and day of each offence, the movement of the victim and offender pre- and post-offence, while drawing on environmental criminology principles (Zipf, 1950; Cohen and Felson 1979; Brantingham and Brantingham, 1981; Cornish and Clarke, 1986).

Using the results of the geographic profile can be an effective way of prioritising databases of possible offenders and/or guiding investigative strategies such as surveillance tactics and house-to-house enquiries.

6.2 Applying geographic profiling to a series of indecent assaults in Bath, England

In 2005, Avon and Somerset Police requested the assistance of the National Centre for Policing Excellence (now the National Policing Improvement Agency) Crime Operational Support following an increase in the number of sexual assaults in Bath city centre. Due to similarities between seven sexual assaults, it was thought that the same offender may be responsible for a number of the offences. All offences occurred at night between 22:00 and 03:30 hours, with the majority occurring on Fridays and Saturdays, when young females (the offender's chosen target population) were walking home from or between social venues. Although the Forensic Science Service (FSS) were progressing forensic opportunities from the cases, no evidence was available to link the offences at that time – or to implicate an offender. The offences had been identified primarily due to their proximity in time and place and because the offender attempted digital penetration. Typically, the offender wore a balaclava and targeted victims wearing skirts. His *modus operandi* was to grab the victims' vaginal area, but would normally flee if the victim resisted.

6.2.1 Geographic profile and supporting analysis

A geographic profile was constructed from the seven offences and was submitted to the enquiry. The CGT analysis produced a peak profile which indicated that a 0.14 km^2 area of Bath city centre was likely to contain an anchor point for the offender. This was 6% of the total hunting area. The results of the CGT analysis were reported with consideration to the crime factors and environmental features of this series. (Operationally, this is a typical process undertaken within the UK when a geographic profile is constructed – contrary to the beliefs of several researchers! (for example see Van der Kamp and van Koppen (2007)). This analysis included an examination of the victim's routes, which highlighted the relevance of the peak geographic profile area as all victims were identified as passing through the peak area at some point in their journey (see Figure 6.1).

Although the offences had mostly occurred in residential areas, the victims' routes had begun in the entertainment locus of the city near to the peak area, and although the victims were not aware of being followed, it was surmised the offender was following them some distance before attacking them. It was likely that he was identifying them somewhere near the peak area and that the victims were defining where the offences would occur by virtue of the route taken on their journey home. The offender would then attack the victim some distance out of the city centre where less surveillance and guardianship was present in the form of pedestrians and CCTV, and would subsequently flee back in the direction of the city centre.

Bath is a substantial entertainment and commercial attractor for the surrounding towns and villages. This together with its university population provides a potentially large pool of victims and could also ensure an offender's behaviour is lost in the anonymity provided by a large town. It was possible that the offender did live within the peak profile area – indeed it contained various multiple occupancy residences, including a bail hostel and YMCA, and just to the north of the peak area was an area of social housing. However, the peak area mainly indicated the central entertainment district within the city centre, including bars, restaurants, shops and clubs. It was therefore feasible that the offender was either working or socialising in this area, commuting in for these activities.

Research conducted by Davies and Dale (1995) found that rapists may travel longer distances to areas where relatively large numbers of potential victims were available rather than offending in the vicinity of their home base. This offender's behaviour, coupled with the demographic attributes of the area was reason for the investigation to focus their proactive resources within the peak, but spread intelligence searches wider as it was possible that the offender was using the centre of Bath as a 'fishing hole'.

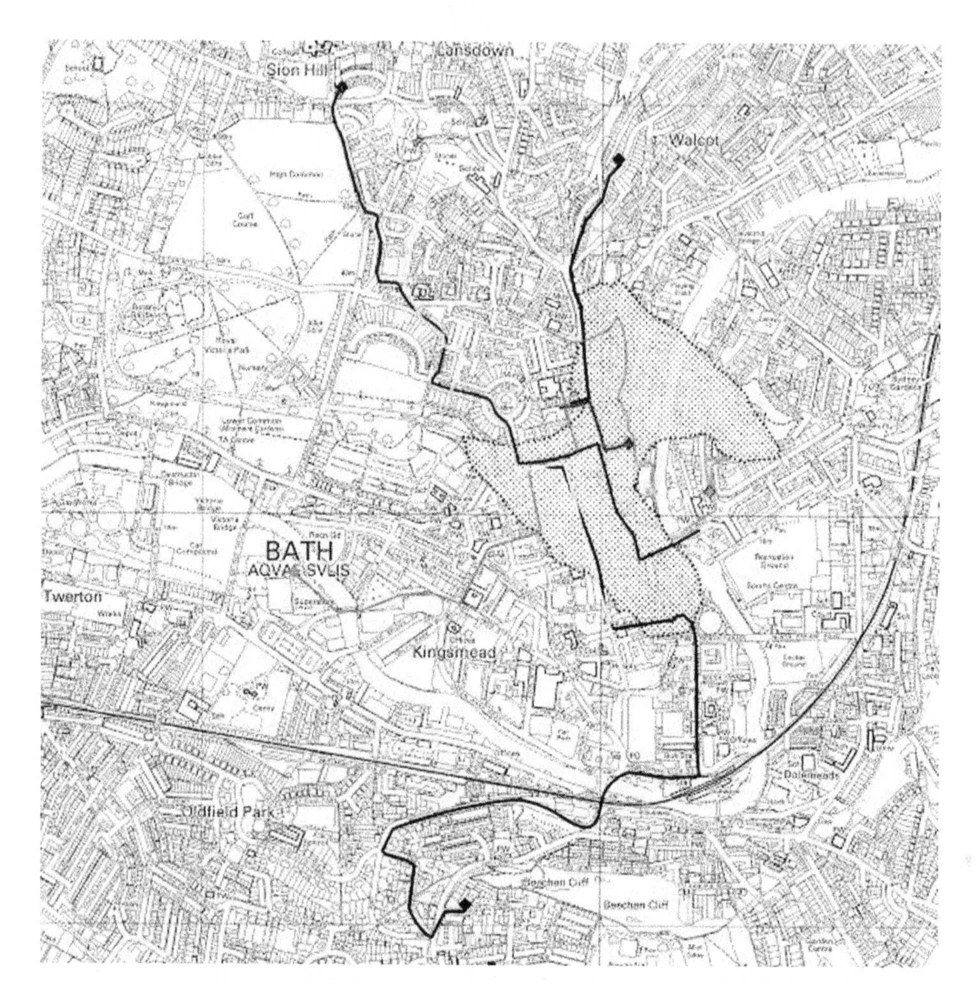

Figure 6.1 The map shows the routes (lines) taken by the seven victims of this sexual assault series, with the peak geographic profile area (the shaded area in the centre of the figure). This shows that all seven victims travelled through the peak profile area at some time on the night they were assaulted. A full-colour version of this figure appears in the plate section of this book.

6.2.2 Investigative strategies

Utilising the peak area, intelligence-based strategies were suggested that focused on identifying residents and employees within the peak profile. Furthermore, due to the times when the offences were occurring and the movements of the victims, pro-active investigative techniques were also focused on the peak profile area. This latter approach was likely to be more productive because the analysis could not for certain indicate whether the peak profile area covered the offender's anchor point, only that he was targeting the area to search for potential victims.

A temporal analysis was conducted, highlighting that the offender was most active between 00:00 and 02:00 hours on Friday and Saturday nights. It was therefore suggested that surveillance activities would be most effective during these times within the peak profile area. In consultation with local investigators who provided knowledge about Bath, a number of residences were highlighted in the peak profile area as worthy of further surveillance and research on the residents. A review of stop-checks and other incidents within the peak area was also suggested in case further incidents were relevant to the series, or indeed whether this could identify possible suspects. The city centre, including the peak area was also covered by CCTV, allowing for operators to monitor the area and remain vigilant with regard to this offender's description and *modus operandi*.

6.2.3 The ongoing investigation

The offender remained active after the production of the first geographic profile. Four months following this geographic profile a number of other offences were under consideration as part of the series, requiring a review of the initial report. Due to escalating factors, a Behavioural Investigative Advisor (a BIA is the UK term for 'offender profilers', who assist investigations by predicting offender characteristics and assessing crime scenes) was employed to assist with series linkage and suggest potential offender characteristics. This linkage analysis cast doubt on the inclusion of two of the offences previously considered as part of the series, and included two later offences that had occurred following the submission of the initial geographic profile. These findings were then considered and two further geographic profiles were generated – one for each of the linkage scenarios. Although differences existed (see Figure 6.2), the profiles defined the same few streets as pertinent to the offender's areal searching activity.

6.2.4 Series resolution

A decoy operation was set up. Over 100 officers were deployed in an attempt to identify the offender actively hunting for victims within the peak profile area. A focused route was planned for the decoy, which incorporated George Street (within the peak of the profile area) and extended out in a northerly direction, similar to the route of a previous victim. At 03.00 hours a male was seen driving slowly around the area, watching and following the decoy to the north of the peak profile area. He watched her disappear into the safe house, then parked, and tried the door of the

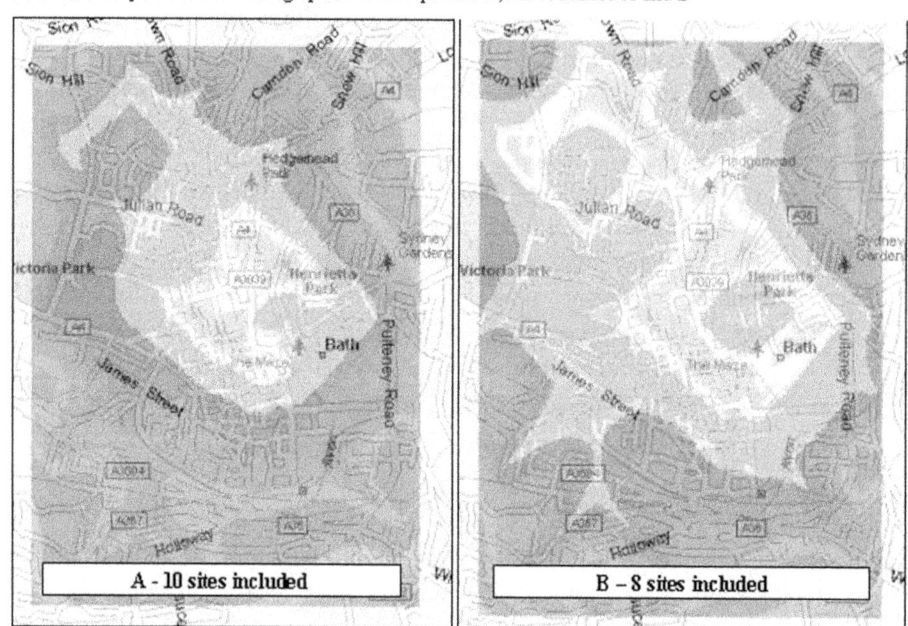

Figure 6.2 The two geographic profiles that were generated as a result of considering (a) ten offences and (b) eight offences. A full-colour version of this figure appears in the plate section of this book.

residence. He was challenged by officers, arrested, made a full admission to nine of the assaults, was sentenced to 9 years imprisonment and is now registered on the sex offender's register indefinitely.

This offender was an example of commuter offenders – a person who travels into the area prior to committing offences. In this case, the peak profile area identified the area the offender was travelling into to initiate their search for victims. Strangeland (2005) documents a similar example.

6.3 Offender geography

As true with any criminal selecting a crime site, the offender in this instance had picked Bath for a reason. Why Bath and not any other location? In our daily routines we most often travel to nearby locations, and perform most activities in areas where we have at least some familiarity. Criminals are no different – after all a target cannot be chosen unless an offender is aware of it (Rossmo, 2000). It was therefore surmised that the offender would have some connection with Bath – be it past or present – that provided knowledge of, and hence a familiarity and sense of comfort with the area in which to offend, even if he was not resident at the time of the offences.

The offender's residence in this series was over 10 km from the peak profile area. He did, however, have five previous residences within the Bath area where he lived for nearly ten years. These probably provided him with the familiarity and comfort with the city centre which resulted in him choosing that location to target victims. The small town centre where he lived and worked may not have provided him with the anonymity required to commit such attacks. Two of the offender's previous addresses were less than 400 m from the peak profile area.

6.4 Operational versus academic geographic profiling

There are a number of ways in which operational profiling differs from the ways in which it has recently been discussed and critically reviewed in the academic press. This has included research that has failed to recognise certain assumptions and criteria that are required for geographic profiling, including:

- the number of events in a series that are required (Rossmo (2000) advises on a minimum of five crime sites to generate a valid and reliable analysis of the offender's potential anchor point);
- accurate identification and linkage of connected incidents;
- exact identification of offence locations;
- awareness of additional spatial information, e.g. offender movement, property deposition;
- knowledge of the overall offence environment.

Although geographic profiling can be flexible to consider fewer than the recommended crimes in a series, results are potentially prone to error.

Geographical profilers utilising geographic profiling software have undergone extensive training and a period of mentorship which ensures they have an understanding of the software system, the methodology behind it and the underlying principles and theories on which geographic profiling is based. The decision to use geographic profiling software in an operational case, and the interpretation of the findings demands that the user is armed with this knowledge. Additionally, many academic evaluation studies make no reference to the training of geographic profilers and have selected serial cases to use in their studies without qualifying whether the case attributes mean they are applicable for geographic profiling.

It has been suggested that in some cases, Rigel (the geographic profiling software used by the NPIA) and other computer systems are no more accurate at predicting the home address of an offender than university students given simple 'rules' which assist their predictions (Snook *et al.*, 2004). This would be expected in some cases, however, these samples have probably included some cases where the utilised heuristics are not deemed applicable. Rigel would not be advocated by a geographic profiling practitioner in these cases and a more in-depth analysis of the offences would be required to understand and explore potential anchor point locations.

This case study demonstrates that in advising an enquiry, consideration is given to a number of factors, such as victim route analysis and environmental criminology principles, in order to develop an understanding of the offender's movements and spatial preferences. This combination of analytical approaches allowed hypotheses to be devised regarding the offender's activities and possible residence.

Academic reference has also been made to geographic profiling's neglect of influencing factors such as opportunity structure, target attractiveness (Bernasco, 2005) and the geographical context of the offences (Strangeland, 2005). However, although these factors are currently omitted from the calculations performed by any algorithm, they are at the core of any operational interpretation and analysis regarding the offence under scrutiny.

Geographic profiling evaluation methodologies have examined the success of the technology and the accuracy of a hit score, with little reference to the intricacies of the temporal and demographic information surrounding a series of offences, or the application of the analytical results to that particular investigation. The processes and considerations involved in the production of a geographic profile go beyond the plotting of points, pressing a button and forwarding the image to the investigator showing a peak area in which they should focus enquiries. Geographic profilers consider: What does the peak area actually represent in real terms? Could it be a residence? Could it be a workplace? Could it be a current or previous anchor point? As well as considering more general principles regarding the offender's spatial activities; Why was the offender in that location at that time? Why did the offender choose that location over other possible locations? If the case study presented here was used in an aggregate evaluation study of geographic profiling it would show Rigel to be inaccurate in predicting the offender's residence.

Operationally the crux of geographic profiling involves combining geographical theory and research with our experience of offender behaviour, while considering relative environmental factors and forensic issues, in order to assess criminal events and guide investigative strategy.

6.5 Conclusions

A geographic profile does not directly solve the crime, but where sufficient spatial and temporal information is available, geographical analysis can help ensure that investigative resources are applied effectively to the point of assisting in the detection of a crime series.

Any geographic profile or geographical analysis will be effective only if the profiler is fully cognisant of all of the case details and investigative avenues. Geographic profiles are calculated for the utilisation of live investigations, to focus resources and prioritise enquiries, and must adapt to the changing needs and faces of that enquiry. The intricacies of live investigations, and the lack of a control situation, make an evaluation of any investigative tool problematic. Indeed, the number of factors involved in the solvability of a crime series is vast. The UK's NPIA has developed a corpus of knowledge, practice guidance and specialised skills in order that Senior Investigating Officers can use the most effective techniques, in the right scenarios to produce the best results. Geographic profiling remains just one of those techniques and tools available to investigators, which if applied at the right time (and to the right case), in combination with other professional practices (such as behavioural advice), can guide an investigation's path. The recommendations it gives do not solve the crimes, they provide an officer with an additional perspective, which should be combined with other analysis and information and used to base decisions in relation to prioritising enquiries.

6.6 References

Bernasco, W. (2005) The use of opportunity structures in geographic profiling. (online) Presentation given at the *Third National Crime Mapping Conference* http://www.jdi.ucl.ac.uk/downloads/conferences/third_nat_map_conf/wim_bernasco.pdf

Brantingham, P. J. and Brantingham, P. L. (1981) *Environmental Criminology*. Waveland Press: Prospect Heights.

Cohen, L. E. and Felson, M. (1979) Social change and crime rate trends: a routine activity approach. *American Sociological Review* **44**: 588–608.

Cornish, D. B. and Clarke, R. V. (Eds) (1986) *The Reasoning Criminal: Rational Choice Perspectives on Offending*. New York: Springer-Verlag.

Davies, A. and Dale, A. (1995) *Locating the Stranger Rapist*. Special Interest Series: Paper 3. London: Police Research Group, Home Office Police Department.

Rossmo, D. K. (2000). *Geographic Profiling*. Boca Raton, FL: CRC Press.

Snook, B., Taylor, P. J. and Bennell, C. (2004) Geographic profiling: the fast, frugal and accurate way. *Applied Cognitive Psychology* **18**: 105–121.

Strangeland, P. (2005) Catching a serial rapist: hits and misses in criminal profiling. *Police Practice and Research* **6**(5): 453–469.

van der Kemp, J. J. and van Koppen, P. J. (2007) Fine tuning geographic profiling. In *Criminal Profiling: International Theory, Research and Practice*, Kocsis, R.N. (Ed.), pp. 347–364.

Zipf, G. (1950) *The Principle of Least Effort*. Reading, MA: Addison Wesley.

7 The Hammer Gang: an exercise in the spatial analysis of an armed robbery series using the probability grid method

Chris Overall and Gregory Day

7.1 Introduction

Primary investigative actions are traditionally focused on the offence location in terms of its value as the foremost source of forensic evidence, victims and/or eye witnesses (Prinsloo, 1996), rather than what it might tell detectives about the spatial behaviour of offenders (Rossmo, 2000). The absence of forensic evidence or eye witnesses at an offence location in a crime series does not necessarily equate to an investigative dead end, but rather the offence location should be seen as tangible evidence in locating the offender by spatially focusing the investigation (Le Beau, 1985). However, the importance of place is often understood by investigators as an indication that the offender is local and who may be confined to a certain neighbourhood, i.e. a few square kilometres, or local as in a municipality or town, which may cover a few hundred square kilometres. Investigative actions that attempt to locate the offender are typically based on the knowledge and inference domains of detectives. The question remains though on whether the inferences drawn by detectives on where the offender may strike next or be apprehended (based on previous experience or knowledge) are sufficiently reliable to be of value to an investigation (Adhami and Brown, 1996; Canter and Alison, 1999).

Information such as land use, area geography, road networks and other infrastructure data, which traditionally may have been viewed as innocuous or irrelevant,

Crime Mapping Case Studies: Practice and Research Edited by Spencer Chainey and Lisa Tompson

takes on a new dimension when integrated with case information and research on the spatial behaviour of offenders (Rossmo, 1995).

The integration of case and geographical data in geographical information systems (GIS) allows for the identification of patterns and relationships between these different types of data, which previously may not have been identified or understood by investigators as being relevant to an investigation (Veenendaal and Houweling, 2000; Canter, 2000). On the basis of this integration, the interception of an offender at a probable future offence location may also present a more cost-effective and alternate investigative option (Helms, 2002) considering that the unknown offender may not always be identified through the criminal justice system (Skogan and Antunes, 1979).

Analysis of serial offender spatial behaviour in South Africa has largely been confined to the serial killer and serial rapist, using GIS to provide investigative assistance in mapping offender movement. This has included the use of Rigel geographic profiling software to determine the home base of the offender (Cooper *et al.*, 2001). No analysis of serial offender spatial behaviour has been completed to date in South Africa using the probability grid method (PGM) (Hill, 2001) to predict the next probable offence location in a crime series such as armed robbery or to test its potential as an investigative decision support tool.

The use of the PGM by Hill (2001) in determining the next probable offence location in an armed robbery series (the 'Video Bandit' series) in Phoenix–Glendale, Arizona illustrated the potential of the PGM as a decision support tool for investigators of serial violent crime. Our objective in this analysis was to test the processes used by Hill in the 'Video Bandit' series to see if similar results could be obtained using a data set of solved armed robberies from South Africa, and be of investigative value to the detectives concerned.

In this analysis, investigative value was broadly defined by the authors as being able to identify the top four potential targets out of a possible twelve in the said data set. This would generate a much reduced geographical operational area and allow for a more cost-effective and practical application of policing resources to focus on prioritised targets. The successful application of the PGM as an investigative decision support tool for detectives in South Africa was potentially far reaching in terms of not only contributing to the reduction of costs and time frames, but also case clearance in serial violent crime investigations (Redpath, 2002).

7.2 Background

A series of seven armed robberies took place in the city of Durban, Kwa Zulu Natal province, South Africa between October 2003 and January 2004. Intelligence suggested that the offender group had committed crime in other areas too, including Gauteng, Free State and Western Cape provinces. The offender group were dubbed

'The Hammer Gang' by the Durban newspaper media on account of a distinct *modus operandi* used in the Durban armed robbery series.

The 'Hammer Gang' would target a particular chain of foreign exchange bureaus where two offenders, one armed and the other using a 5 kg hammer, would smash the bullet-proof glass of the teller's booth, threaten staff with a firearm and remove cash. Another two armed offenders would wait outside the bureau to prevent any interference from mall security guards. The offenders would then leave the hammer behind and flee the scene in a waiting getaway vehicle. All of the armed robberies took place in suburban shopping centres in Durban, close to major arterial roads. The extent of the offenders' spatial activities ranged over four different policing operational areas, totalling 148 km^2.

Due to the nature of the offences, this armed robbery series was investigated by the Provincial Serious and Violent Crimes Unit (this group has now been restructured to fall under the banner of Organised Crime in the South African Police Services), Kwa Zulu Natal of the South African Police Service and the offenders were arrested soon after the last offence in January 2004. This post hoc analysis was undertaken by the authors in January 2006 once this data set had been made available to them by the investigating officer.

7.3 Mapping the data and getting the picture

Given the definitive type of the offender targets, the first step was to map all twelve locations of this particular chain of foreign exchange bureaus in the Durban area and overlay them with six of the seven offence locations. This was done using a simple data input script developed by the authors in ArcGIS. Mapping the offence locations gave us a clearer understanding of the extent of the offenders' spatial activities in the series as we could now visualise it against a digital ortho-photography backdrop (see Figure 7.1). This, however, gave no clear indication of 'where to next' in terms of offender movement or how this related to the other potential targets identified.

The offence location data were then imported into an Environmental Systems Research Institute's (ESRI) tactical crime analysis software extension to ArcGIS called 'Crime Analysis Tools". The Crime Path tool was then used to create a sequential crime path from the offence locations. Once this had been done a distinctly circular sequential crime path was formed from the way the previous targets had been selected by the offenders as if to join the dots and close the circle of their activities.

Although the sequential crime path infers the most logical next and last (seventh) offence location, the question is whether or not investigative experience/hunch or logical deduction alone would have been sufficient for the making of an effective tactical investigative decision (Helms, 2000).

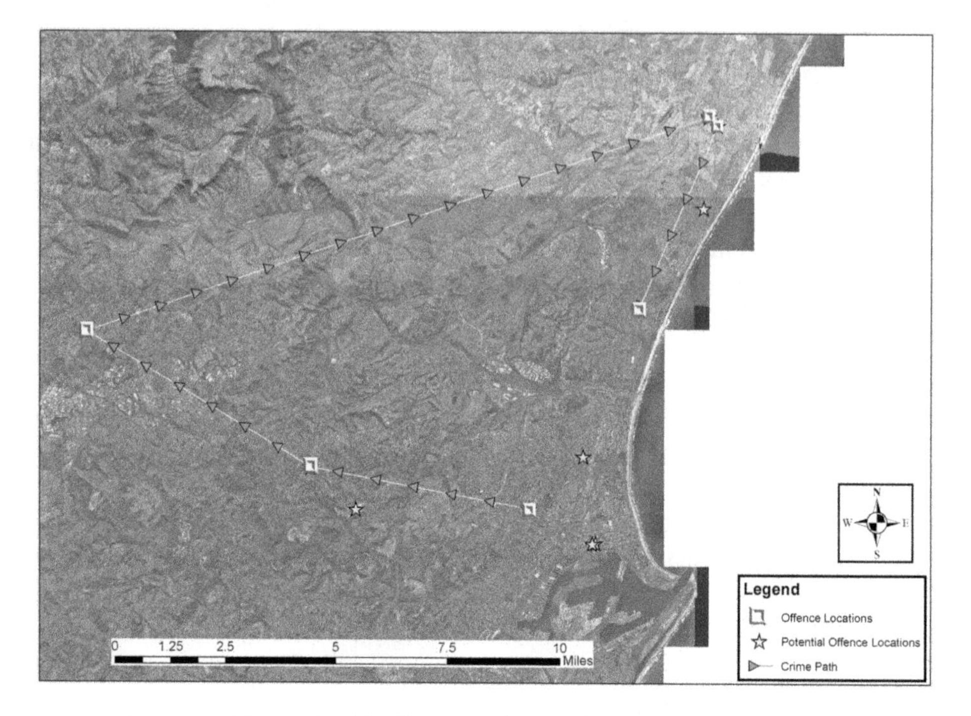

Figure 7.1 Crime path with offence locations and potential offence locations.

7.4 Predicting the next offence location

The first step in the analysis process was to create the probability rectangles, the assumption being that if the previous offences have occurred in the 68th and/or 95th percentile areas then the next offence location should be located in the same 68th and/or 95th percentile areas (see Figure 7.2).

Once this had been carried out it was noted that one of the potential offence locations was located outside the 95th percentile area. Since the area covered by the 65th and 95th percentile areas were 179 and 228 km^2 respectively, the exclusion of one potential offence location still left eleven out of twelve potential offence sites in a very large operational area.

The use of probability rectangles as a target discriminator in this instance would not have been operationally useful for investigators given the manpower and assets needed to address that many targets spread across the city. Similar results were observed by Catalano (2000) and Hill (2001) in their respective application of probability rectangles where offenders operated over large geographical areas.

The next step was to follow the processes outlined by Hill (2001) and create a probability grid in ArcGIS by which accompanying spatial and case data layers

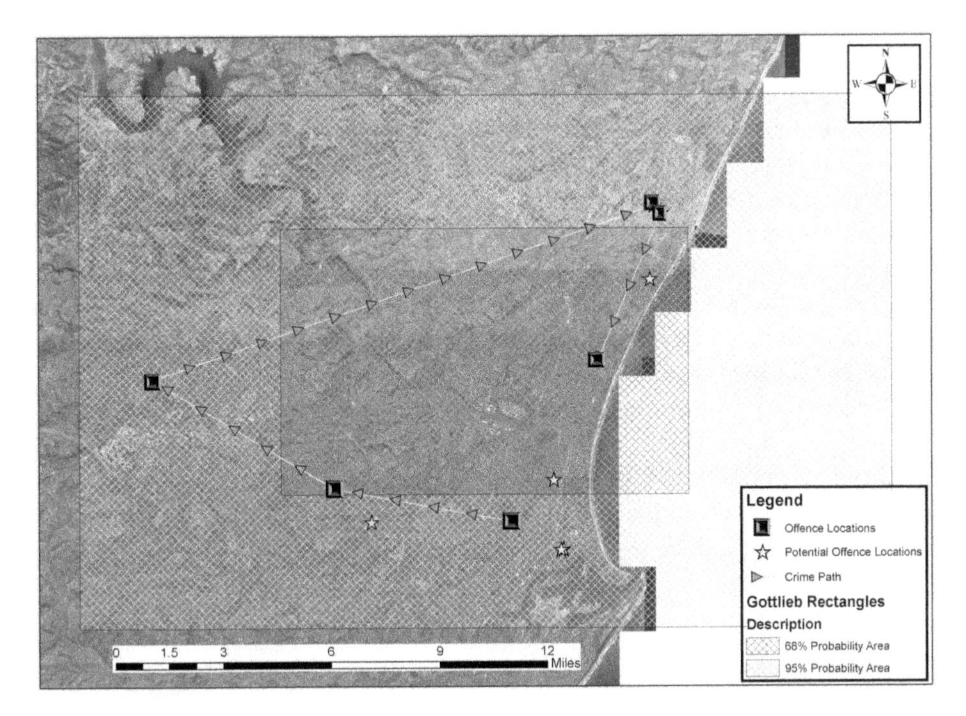

Figure 7.2 Gottlieb rectangles with crime path, offence and potential offence locations.

could be scored and combined into a single data set to produce a highest probability target area. The probability grid cell size was established at 1 km^2 in view of the spatial distribution of existing potential targets. These more refined results helped to reduce the potential target count and investigative operational search area associated with them.

Scoring the probability grid map was based on where the grids were located in relation to the following spatial data layers and scored on whether or not a grid was located:

1. within the Convex Hull Polygon
2. within the Probability Rectangles
3. within a Land Use type
4. within the Last Hit Buffers
5. whether targets were available in a grid
6. direction of crime path

7.5 Results

The resultant probability map produced two very high-probability offence locations in close proximity to each other, one being the seventh and last offence in this series. Based on the mean distance travelled between offences (9.7 km), the logical assumption was that the next offence location would be located within a radius of 9.7 km from the sixth offence location or within the first ring buffer. Both of the very high probability target locations were located within the first ring buffer and the calculated radius was in line with the offenders' directional path. Taking into account the number of potential targets, the PGM reduced the potential target locations from eleven to two (see Figure 7.3).

Reduction of the investigative operational area was substantial to the point that the two very high probability targets were 2.7 km apart and located in the operational precinct of one police station instead of a previous operational area of 228 km^2 containing the precinct areas of four South African Police Service stations.

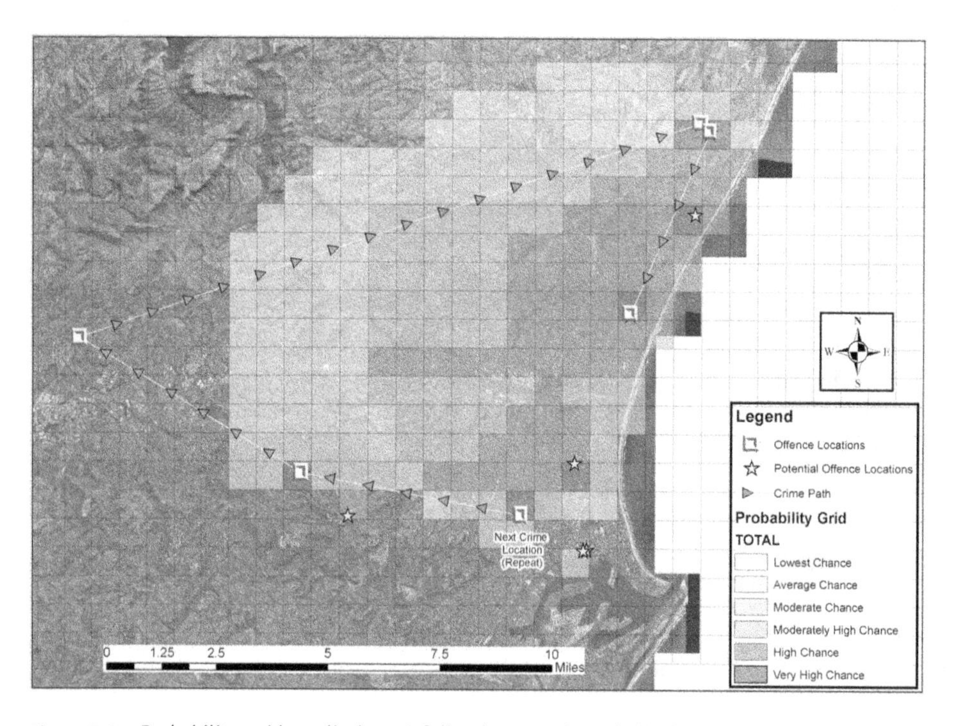

Figure 7.3 Probability grid prediction. A full-colour version of this figure appears in the plate section of this book.

7.6 Issues in application of the probability grid method

The application of the PGM as an investigative decision-support tool in South Africa will inevitably usher in the increased use of mapping technology and a change in the required domain knowledge of analysts and detectives alike. In this respect detectives and analysts may also have to consider socio-economic, geographical and infrastructure variables that traditionally have not been factored into serial crime investigations.

Independent variables such as the impact of media reports, resultant target hardening and increased law enforcement activity in potential target areas will almost certainly influence offender spatial behaviour, whether or not a crime series will continue and the subsequent spatial analysis result.

When carrying out this analysis the authors did not have to factor in the impact of the above independent variables, interpretation of offender movement or intuition/experience of analysts/investigators in determining the value of data and how it should be applied in a live analysis. Although detectives and analysts may use their investigative experience or intuition to interpret offender spatial behaviour, there is little in the way of research into the spatial behaviour of South African offenders to guide this interpretation (e.g. see Oldfield, 2000). Whether current international research on spatial behaviour of offenders can be applied as a guide in the context of South African serial crime investigations remains to be tested. The authors in this instance did refer to current international research on the spatial behaviour of offenders as a guide in the spatial analysis process.

7.7 Conclusions

'The Hammer Gang' offenders provided a virtual text book example of spatial behaviour in terms of consistency and target specificity which allowed for a relatively simple analysis process to be carried out. As an investigative decision-support tool, the PGM provided a valuable investigative result by identifying the next offence location correctly and greatly reducing the geographical operational area. However, the same result and level of analytical complexity may certainly not be true for crime series where offenders are not spatially or target consistent. Both Catalano (2000) and Hill (2001) in their respective papers on the use of the PGM in crime series have indicated that limitations exist in this application and that is something that we will need to be mindful of in future application of the PGM. A true evaluation of PGM as an effective investigative decision-support tool will see this application being tested across a range of different crime type series in South Africa in coming times.

7.8 Acknowledgements

The authors wish to thank Director Johan Booysen, Provincial Head of Organised Crime, Kwa Zulu Natal, South African Police Services for facilitating access to the case data and making this analysis possible.

7.9 References

Adhami, E. and Brown, D. (1996) *Major Crime Enquiries: Improving Expert Support for Detectives*. London: Special Interest Series Paper 9, Police Research Group, Home Office.

Canter, D. and Alison, L. (1999) Professional, legal and ethical issues in offender profiling. In *Profiling in Practice and Policy*, Canter, D. V. and Alison, L. J. (Eds), pp. 23–54. Liverpool: Offender Profiling Series Vol. 2, Centre for Investigative Psychology.

Canter, P. (2000) Using a geographical information system for tactical crime analysis. In *Analysing Crime Patterns: Frontiers of Practice*, Goldsmith, V., McGuire, P., Mollenkopf, J. and Ross, T. A. (Eds), pp. 3–10. London: Sage Publications.

Catalano, P. (2000) Applying geographical analysis to serial crime investigations to predict the location of future targets and determine offender residence. Presented at *Crime Mapping: Adding Value to Crime Prevention and Control Conference*, 21–22 September, Adelaide, Australia. Australian Mineral Foundation.

Cooper, A., Byleveld, P. and Schmitz, P.M.U. (2001) Using GIS to reconcile crime scenes. Presented at the *Fifth Annual International Crime Mapping Research Conference*, 1–4 December, Dallas, TX.

Helms, D. (2000) *The Use of Dynamic Spatio-Temporal Analytical Techniques to Resolve Emergent Crime Series* (online). International Association of Crime Analysts. http://www.iaca.net.Articles/dynamic.pdf

Helms, D. (2002) The tactical checklist. In *Advanced Crime Mapping Topics*, Bair, S., Boba, R., Fritz, N., Helms, D. and Hick, S. (Eds), pp. 7–21. Denver, CO: NLECTC, University of Denver.

Hill, B. (2001) Narrowing the search: utilizing a probability grid in tactical analysis. Presented at *The Fifth Annual International Crime Mapping Research Conference*, 1–4 December, Dallas, TX.

Le Beau, J. L. (1985) Some problems with measuring and describing rape presented by the serial offender. *Justice Quarterly* **2**: 385–398.

Oldfield, D. (2000) What help do the police need with their enquiries? In *Offender Profiling: Theory, Research and Practice*, Bekerian, D. and Jackson, J. (Eds). Chichester: J. Wiley & Sons.

Prinsloo, J. (1996) The scene of the crime as a source of information. In *Forensic Criminalistics*, 2nd edn, Van der Westhuizen, J. (Ed.). Johannesburg: Heinemann.

Redpath, J. (2002) *Leaner and Meaner? Restructuring the Detective Service*. Pretoria: Monograph 73, Institute for Security Studies.

Rossmo, D. K. (1995) Place, space, and police investigations: hunting serial violent criminals. In *Crime and Place: Crime Prevention Studies*, Vol. 4, Eck, J. E. and Weisburd, D. A. (Eds), pp. 217–235. Monsey, NY: Criminal Justice Press.

Rossmo, K. (2000) *Geographic Profiling*. Boca Raton, FL: CRC Press,

Skogan, W. G. and Antunes, G. E. (1979) Information, apprehension and deterrence: Exploring the limits of police productivity. *Journal of Criminal Justice* **7**: 217–241.

Veenendaal, B. and Houweling, T. (2000) Gut feelings, crime data and GIS. Presented at *Crime Mapping: Adding Value to Crime Prevention and Control Conference*, 21–22 September, Adelaide, Australi. Australian Mineral Foundation.

8 'Rolling the Dice': the arrest of Roosevelt Erving in Lincoln, Nebraska

Tom Casady

8.1 Introduction

Geographical analysis enriches police decision-making in a variety of situations. Among its uses is helping to design effective investigative strategies. The 14 January 2005 arrest of a serial bank robber in Lincoln, Nebraska, USA is an excellent example of how good analysis augments good police work.

About 16:00 hours on 6 July 2000, a man wearing a blue bandana over his face walked into a branch office of U.S. Bank, located at the end of a small neighbourhood shopping area at 13th and Arapahoe Streets in Lincoln. A robbery was underway. The suspect brandished a semi-automatic pistol, threatening the three employees and demanding the cash drawer contents. He ran out the door with a bag filled with $14,406. A witness in the parking lot saw the suspect run to a green Ford pickup truck and drive away. This vehicle was found abandoned about six blocks away shortly after the robbery: it had been stolen earlier in the day from the owner's driveway. Thus began a series of robberies that frustrated Lincoln police officers for the next four years.

The series ultimately ended on the snowy morning of 14 January 2005. At 07:51 hours that morning, Officer Scott Arnold interrupted a robbery in progress at a Wells Fargo Bank branch in east Lincoln. Shortly thereafter a 39-year old suspect, Roosevelt Erving, was arrested for thirteen bank robberies that had netted $445,203.

Crime Mapping Case Studies: Practice and Research Edited by Spencer Chainey and Lisa Tompson

8.2 Erving's series of bank robberies

Bank robberies are comparatively rare in Lincoln, Nebraska's second-largest city, with a population of 241,000. Erving's first robbery would be the sixth of nine bank robberies to occur in the city in 2000. For the past ten years, the city has averaged just six bank robberies each year, so 2000 stands out as a particularly high year. Erving's second robbery occurred four months later, on 4 November 2000 at another suburban branch bank. The take was much smaller: $7,744. Six weeks later, he struck again, returning to the U.S. Bank branch where he had started, and making off with almost as much money. By the time of the third robbery, Lincoln police were certain that a single suspect had committed all three. The *modus operandi* was clear, and remained similar for the first seven robberies. Erving would steal a vehicle to use in his get away, and abandon the vehicle nearby shortly after the robbery. In each case, Erving brandished a pistol, covered his face with a bandana, and demanded cash from the teller line. Unlike his first robbery, which was in the late afternoon, he committed the next four in the middle of the day.

In his sixth robbery, on 19 June 2001, Erving discovered the vault. In this case, after clearing the teller drawers, Erving ordered the two employees into the vault, gathered more cash and forced them to lie on the floor. It was his biggest score to date: $66,200. Erving was learning the ropes of becoming a better bank robber.

Robbery number nine was his biggest haul. At about 07:00 hours he abducted an employee arriving for work in the parking lot at gunpoint, waited for a second arriving employee, then forced both into the bank and into the vault. He again made the employees lie on the vault floor. Erving left with $94,859, and had graduated to a new and more lucrative method: the morning glory robbery. Hitting banks at opening time is an old *modus operandi*, but Erving now had uncovered the peculiar effectiveness that has made the morning glory robbery method popular. The remaining five robberies of his career all followed this same pattern.

As Erving's bank robbery career developed, the stakes raised considerably. As his *modus operandi* developed and changed, he was not only making more money, he was also terrorising employees much more. Rather than brandishing a pistol and leaving quickly, he was abducting employees, staying longer, forcing them into other parts of the bank (the vault) at gunpoint, and making them lie on the floor. At his tenth robbery, on 20 January 2004, he tied the employees up.

It is hard to understand the terrorising effect the later robberies had on the employees without understanding the current events of the time. Nebraska, although geographically large, has a population of only 1.7 million. It has a small media market, dominated by a handful of daily newspapers, news radio stations and television stations. In September 2002, a U.S. Bank branch in Norfolk, Nebraska had been robbed. All the employees – five in total – had been executed by the robbers. These men had been captured, and the cases were proceeding during the ensuing two years. This story was dominating Nebraska news across 500 miles. The victims of Erving who were tied up and told to lie on the floor in the vault thought they would die.

The Lincoln Police Department and the Federal Bureau of Investigation had been working hard to identify the serial bank robber, and to make an arrest, but nothing had panned out. Unlike many serial bank robbers, this suspect did not seem to be surfacing in other cities. The efforts – although extensive – had resulted in very few credible leads. One investigator, though, had made a cogent observation: Detective Sergeant Jim Breen of the Lincoln Police Department had noted that the time period between the robberies appeared to be loosely related to the amount of money the suspect obtained at each.

8.3 Analysing Erving's series

On Tuesday, 14 December 2004, the command staff of the Lincoln Police Department held its regular weekly meeting. That morning, the group discussed the bank robbery series. It had been six weeks since the last robbery, where the suspect had made $44,000. At the meeting, the group decided that the department had to do something more, and a decision was made to implement a stakeout project.

Stakeouts are an expensive gamble. The logistics are substantial, the cost – especially personnel costs – are high, and the chances of success are poor. Virtually all experienced cops have burned the midnight oil on a stakeout at one time or another, only to have the criminal commit another offence shortly after the stakeout ends. The department, however, could not simply wait for the next robbery. Assistant Chief John Becker was tasked with assembling the plan, and the project was dubbed 'Rolling the Dice', a reference to the understanding that this was a low-percentage gamble.

Good analysis, however, was about to swing the odds in favour of the police. One of the first steps in the analysis was to pick up on Detective Sergeant Breen's observation. Placing the robberies and amounts on a timeline, the analytical team calculated the days between each robbery and the amount of money the suspect took. The data confirmed that the suspect would be likely to strike again in the early part of 2005. The pattern, however, was not so consistent as to narrow the time period down precisely. The police department could not afford to run this project very long – the manpower needs and costs were going to be high in a lean budget year. It was decided that eight days was the maximum the department could afford. Although the financial analysis resulted in a rather broad time window during the first quarter of the year, the planning team settled on the middle of January 2005 as the earliest likely time period for the next robbery in the series.

In preparation for the project, a considerable amount of analysis was done to identify the most likely targets. First, geographical analysis revealed that all of the robberies had occurred in the southeast area of the city, with one exception. The single exception was just a few blocks out of the southeast quadrant. Moreover, the robberies had occurred close to the city's edge. There were also distinctive characteristics to the types of banks that were robbed. Erving had returned to the same branch banks on three occasions, accounting for six of the 12 robberies. All of the branch banks were small, freestanding branches with few employees. None were located in buildings containing other offices. It was also fairly clear that after the morning glory pattern emerged, Erving was casing the banks: he knew when the employees would be arriving, what they would be driving and how many employees would be opening.

Armed with this information, analysts produced maps of all Lincoln's financial institutions, highlighting approximately 35 branch banks that met the profile. The stakeout detail of 30 officers and four FBI agents would cover 12 of these locations with a two-person team, with a reserve force of SWAT Team members ready to respond in the event of an actual robbery. Mapping revealed that a stakeout team could effectively cover multiple branch banks at certain locations. The remaining likely targets would be immunised in one of two ways: either by parking empty marked police patrol cars nearby in the days preceding the stakeout, or by directing the on duty officers in marked units to patrol the approaches to that bank during the opening time period from about 07:00 to 8:30 hours on those days. The plan was to funnel the robber to a small number of potential targets.

Analysts had also examined potential escape routes from the southeast quadrant of the city. Since the prevailing theory was that the robber was coming to Lincoln from outside the city, the team made arrangements for potential exit roadways to be kept under surveillance by surrounding law enforcement agencies, including the Nebraska State Patrol and the Lancaster County Sheriff's Office. Deputies and troopers would be stationed at key locations to observe outbound traffic in the event that a suspect vehicle description materialised.

8.4 Project 'Rolling the Dice'

The project was set to begin on 14 January 2005. This was a Friday before a three day weekend, with the Martin Luther King Day national holiday on Monday 17 January. The team felt that a three-day holiday weekend might be a time when the robber was anxious to stock up on cash if he was using the proceeds to fund gambling, drugs or other leisure activities.

The first day of the stakeout project began with a 04:00 hours briefing in the assembly room at police headquarters. The entire team was brought up to speed, the plan reviewed and the specific assignments made. The detail was to deploy and be operational at 06:00 hours.

Detective Sergeant Dennis Duckworth was tasked with staffing the command post for the project, located in the city's emergency operations centre. Sergeant Duckworth had prepared a dry-erase board to record times and remarks concerning key events. Entry number one, at the top of the board was: '0400: Briefing at HQ'. Entry number two was: '0600: All units in post'. At 07:51 hours, 31-year veteran patrol officer Scott Arnold was driving his marked patrol car through the parking lot of a Wells Fargo branch bank at 66th and O Street. Officer Arnold was one of the 'vaccinators' – watching closely a few branches on his beat that were not covered by the stakeout teams. He noticed something peculiar; the bank's drive through lane was not yet open despite the employee's cars in the lot and an expected 07:30 hours opening time. Further observations caused him to suspect that a bank robbery was in progress. He summoned assistance to set up a perimeter.

He was right. Roosevelt Erving had arrived at the bank shortly before opening time, and abducted a teller arriving for work. He was armed with a Glock 9 mm semi-automatic pistol. He had taken her inside at gunpoint, and waited for the second employee to arrive. Both employees had been restrained to chairs using cable ties. Officer Arnold had indeed interrupted a robbery in progress, and broke the radio silence that had been part of the detail. Erving, however, saw the patrol car, and quickly realised that the police had discovered the robbery. In an effort to avoid the perimeter officers that were arriving on the north, he threw a chair through a south-facing window on a side of the building with no exits. Officer Arnold spotted him, and gave chase on foot.

Erving crossed O Street, Lincoln's busiest seven-lane street. The pursuit went through the parking lot of a restaurant on the southwest corner of 66th and O Streets, where coincidentally, Michael Heavican, the United States Attorney for the District of Nebraska, Paul LaCotti, Special Agent in Charge of the Nebraska FBI office, and Colonel Tom Nesbitt, the Superintendent of the Nebraska State Patrol, were having breakfast.

The pursuit continued between buildings, through the parking lot of a retirement home, and then between houses into a residential neighbourhood. As the events had unfolded, Sergeant Ken Koziol a 30-year veteran, had left his position on the stakeout team at a Union Bank branch nearby, and made his way into the neighbourhood south of the chase. He spotted Erving, with Officer Arnold following, and took up a position to intercept. As Erving went up the walkway of a residence, Sergeant Koziol came around the corner, shotgun in hand and confronted Erving. Officer Arnold arrived, and handcuffed the suspect.

Entries number three and four on Sergeant Duckworth's log: '0751: Robbery in progress; 0807: Suspect in custody'.

Erving had discarded his pistol during the foot pursuit, and it was recovered a few days later, as the snow was melting. It turned out that he was a local – he lived in the southeast part of Lincoln, and had committed the robberies after getting off his night shift job. Erving had purchased a home and several vehicles with the proceeds of his crime spree. In December 2005 he was convicted in Federal court to thirteen counts of bank robbery, and is presently serving a 40-year sentence.

8.5 The crucial role of geographical analysis

There can be little doubt that a good deal of luck was involved in the arrest of Roosevelt Erving. It is also undeniable, however, that good analysis played a crucial role. Crime mapping helped determine Erving's *modus operandi*, helped locate his potential victim banks, helped determine the best locations for stakeout and helped identify the likely escape routes. The case is illustrative of the nexus of the art and science of policing: the creativity, initiative and intuition of experienced police officers informed by geographical crime analysis and traditional investigative analysis. The combination certainly created one of the most memorable moments for two veteran police officers who rolled the dice and hit the jackpot.

Part III Neighbourhood analysis

9 The strategic allocation of resources to effectively implement Neighbourhood Policing and the Community Safety Plan

Alice O'Neill

9.1 Introduction

The National Community Safety Plan 2006/9 (Home Office, 2005) showed clear direction for the police service in two ways.

- *Neighbourhood Policing* – local priorities identified by local people as the issues that need to be dealt with in their area with police and partners working together to tackle them.
- *Managing protective services* – such as counter terrorism and tackling serious and organised criminality.

West Midlands Police (WMP) realised that these potentially could be conflicting 'pulls' in terms of resource allocation and prioritisation of threats. It was therefore imperative to take a fresh approach to the Police Force's Strategic Assessment – the Strategic Assessment is one of four types of 'intelligence products' produced under the England and Wales National Intelligence Model (for more information see http://www.police.uk/nim2) – and importantly the threat assessment to ensure that the Force was effectively sighted and could deliver against these competing demands.

Crime Mapping Case Studies: Practice and Research Edited by Spencer Chainey and Lisa Tompson

The existing threat assessment was not considered fit for purpose, nor flexible enough to cater for the different priorities of the Police's Basic Command Unit area (BCUs) and neighbourhoods. It was also considered to be inflexible to different National Intelligence Model (NIM) levels across the Force.

In response to this, WMP developed a Strategic Threat and Risk Assessment Index (STRATi), designed to have an impact on three key areas which would deliver the Chief Constable's vision; 'To reduce crime and disorder and make our communities feel safer'. The three core areas that were assessed were:

- *Protecting the public* – protecting the public from death and/or serious injury, both physical and psychological.
- *Promoting community stability* – promoting community stability and reassurance.
- *Reducing victimisation* – leveraging opportunities against the Force's crime reduction target of a 15% reduction in total recorded crime (in England and Wales this is commonly known as 'Public Service Agreement (PSA) target 1').

A weighting factor was also applied to account for which NIM level the index was applied to, enabling the intelligent prioritisation of risk on all NIM levels. This meant that at a Force level (i.e. levels 2 and 3) the area of 'protecting the public' was weighted more heavily because at Force level exists the responsibility and access to resources to tackle serious and organised criminality (such as disrupting an Organised Crime Group involved in cash in transit offences through the use of resources such as the surveillance unit). At a BCU level (level 1) they are more equipped to deal with volume crime such as theft from motor vehicles and therefore 'reducing victimisation' is weighted more heavily. Similarly at Level 0 – the neighbourhood level – it allows the focus of delivering a service against the issues identified as a priority by the local community.

9.2 Alternative resource allocation model

The alternative resource allocation model was an attempt to influence resource allocation to ensure that it was intelligently focused on the type of demand experienced across WMP. The levels of crime and disorder were geographically risk assessed in context of the geography and socio-demographics across the Force. This showed that some areas of the Force had a significantly higher level of some crime types – such as Birmingham city centre records a high level of robbery whereas another BCU has a high volume of anti-social behaviour (ASB) and yet another BCU had a high fear of crime which was not reflected in recorded crime. All of these require a different type of police response. It enabled us to show that different areas had differing issues and concerns were not confined to traditional BCU boundaries and therefore could not be addressed by a 'one size fits all' approach.

A variety of information sources such as crime and incident information were utilised, alongside partner agency information such as Drug Intervention Programme data, deprivation data and census information such as proportion of elderly, proportion of young people and proportion of unemployed persons. We also utilised the 'Feeling the Difference' survey information (a Force funded public perception survey), which contained a raft of questions that can be used as a tool to measure our 'customer satisfaction' and the impact of policing. It asks questions such as 'how satisfied or dissatisfied are you with the level of foot patrol in your neighbourhood?' and 'what is the biggest crime or anti-social behaviour issue in your neighbourhood?' The results of this were analysed and correlations made to actual levels of crime and extrapolated across the Force. This enabled us to be more pre-emptive in identifying areas across the Force that may have similar issues and require a similar response strategy.

This was the foundation for developing an alternative resource allocation model to reflect the change in the policing family. The WMP were planning on introducing approximately 1000 Police Community Support Officers (PCSOs) over the financial year 2006–07 and therefore we wanted to give the decision makers a guide to the areas that would benefit most from these resources. Central to this was the fact that PCSOs are more equipped with the tools to deal with local issues such as anti-social behaviour at the neighbourhood level, rather than serious crime situations where they may not possess the required powers. This would place the Force in a good position to ensure the delivery of neighbourhood policing across WMP while addressing the issues surrounding protective services. Ensuring the right people are in the right numbers, in the right places and, importantly, addressing the right issues and concerns is paramount.

9.3 What were the results, outcome and issues?

One success of this model was to assist BCUs in identifying if they had significant level 2 issues on which they currently had no focus as they were concentrating on volume crime to deliver against PSA 1. For example some BCUs did not fully recognise the impact they had both locally and at a Force level on the criminal use of firearms or Class A drugs misuse – they were not looking at the issue unless it formed part of a Force-led initiative. Identifying these threats to the Senior Management Team and presenting them as a critical risk which needed managing, rather than the previous top-down approach was a significant change. It has encouraged a bottom-up approach to obtain the community intelligence that could feed a level 2 approach, if it was appropriate for the situation.

The model also identified geographical areas that offered leverage against PSA 1, by highlighting areas that experienced a high level of crime that offered potential reduction opportunities in terms of the spread of offending. This enabled the Force, through the Tactical Assessment process to allocate resources accordingly, aiming to maximise the gains from the minimum number of resources.

Utilising a variety of information sources within strategic analysis has enabled WMP to develop a more detailed understanding of reported and unreported crime and also on the perception of crime. This has given the Force a baseline against which to measure performance.

The aim of this model was to take Force resources back to the centre and re-think in what areas those resources would be best situated. Unfortunately, the timing of this proposal coincided with the potential implementation of Police Force mergers to create 'super police Forces'. Although the alternative resource allocation model was not adopted in relation to the allocation of PCSOs it did generate a lot of thought and discussion at a senior level. For example, the use of PCSOs and the Special Constabulary was considered in areas of the Force where they may have the maximum effect to enable the reallocation of police officers based on demand.

9.4 The future

A further development of this model is a STRATi tactical model. From the analysis we know that some crime types increase at certain times of the year; for example personal robbery traditionally increases in the West Midlands between November and March. A tactical version of this model would be able to monitor any unseasonable activity or emerging threats to support resource prioritisation.

Future developments of this model are planned to incorporate partner agencies' critical risks to fully exploit the joined up opportunities required in the National Community Safety Plan. This will enable our partners to easily assess the level of threat posed by different areas – as all levels of threat are assessed by a common methodology. A weighting factor can be applied according to the agencies' priorities which will then identify the areas where there is the most leverage in terms of achieving the objectives within those priorities.

Change is challenging in an organisation as large as WMP and we remain hopeful that this approach did result in 'turning some small cogs in some large minds' and will continue to generate discussion and assist in the allocation of resources against competing demands in the future.

9.5 Reference

Home Office (2005) *The National Community Safety Plan* (online). London: Home Office. http://www.crimereduction.gov.uk/communitysafety01a.pdf

10 Priority neighbourhoods and the Vulnerable Localities Index in Wigan – a strategic partnership approach to crime reduction

Ian Bullen

10.1 Introduction

The Vulnerable Localities Index (VLI) was a police response to measuring community cohesion. Developed jointly by the Central Police Training and Development Authority (Centrex) and the Jill Dando Institute of Crime Science, it followed in the wake of the riots that occurred during 2001 in Oldham, Burnley, Bradford and Wrexham as an attempt to identify priority areas where similar breakdowns in community cohesion might occur (Chainey, 2004).

The index uses a broad range of data to do this, principally due to a need to apply a consistent methodology nationwide. This methodology needed to use data that would be easily accessible and usable (all data is calculated to census Output Area level (OA)), involve little or no training for the analysts expected to implement it, produce concise output for managers and be simple and practical to apply, yet robust (Chainey, 2004).

The process of identifying vulnerable localities is the first stage in a process that follows the 'Scan Analyse Respond Assess' (SARA) cycle familiar in policing circles. It is the start of the process of understanding local problems. Objective statistical information, collated and aggregated to the VLI score, needs to be viewed subjectively by those practitioners with a good local understanding.

Crime Mapping Case Studies: Practice and Research Edited by Spencer Chainey and Lisa Tompson

Six locality statistics are normalised and aggregated to arrive at an overall VLI value for every Census OA in a district. The six statistics are:

- burglary dwelling incidence rate per 1000 households per annum;
- criminal damage to a dwelling incidence rate per 1000 households per annum;
- income deprivation index value (based on five domain indicators looking at Income Support households, income based Job Seekers' Allowance households, certain Working Families Tax Credit and Disabled Person's Tax Credit households, and National Asylum Service Support households);
- employment deprivation index value (based on six domain indicators – counts of unemployment claimants, incapacity benefit claimants, Severe Disability Allowance claimants, and New Deal participants);
- percentage of population with qualifications below level 2 (5+ GCSEs graded A–C);
- percentage of population aged 15–24 years inclusive (Chainey, 2004).

10.2 An alternative Vulnerable Localities Index

Within Wigan Borough there was a collective desire to understand the nature of these statistics and also to examine if the desired 'robustness' could be verified. In order to do this, a further six locality statistics were identified and subjected to the same normalisation and aggregation process to arrive at an overall 'alternative' vulnerable localities index value for every census output area. The 'alternative' statistics selected were:

- violent crime incidence rate (within a domestic setting) per 1000 households per annum;
- Greater Manchester Fire and Rescue Service deliberate/malicious fires incidence rate per 1000 households per annum;
- probation client incidence rate per 1000 households (by last known address);
- Substance Misuse Services client incidence rate per 1000 households (by six Figure postcode of last known address);
- health deprivation index value (based on four domain indicators looking at years of potential life lost, comparative illness and disability ratios, emergency admissions to hospitals and mood or anxiety order amongst adults under 60);
- percentage of lone parents in a household with dependent children.

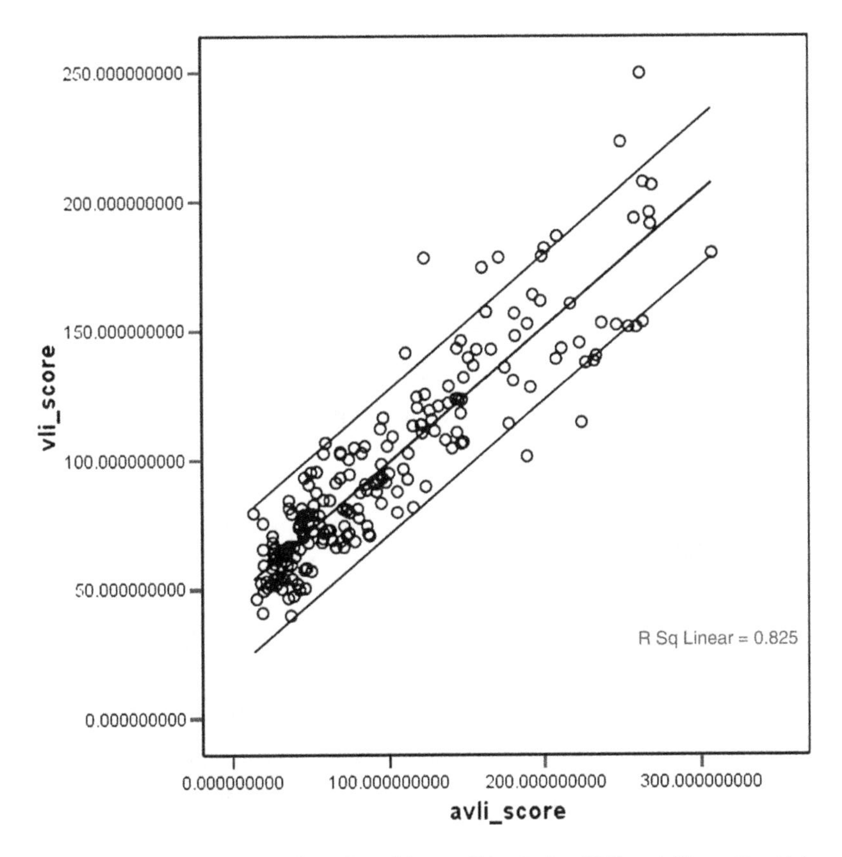

Figure 10.1 Correlation between the Vulnerable Localities Index (VLI) and Alternative Vulnerable Localities Index (AVLI) scores in the borough of Wigan.

The construction of the 'Alternative' Vulnerable Localities Index (AVLI), as it became known, helped us to do two things. First, it showed a correlation with the existing VLI scores (R Sq Linear 0.825 – see Figure 10.1), confirming the robustness of the initial data sets chosen. Second, it provided us with an additional six data sets that readily helped us to understand the nature of, in particular, the most vulnerable areas within the Borough and that added a sharper focus to the scanning process.

The two scores were then combined to produce an aggregate VLI score and this score was then used to identify the most 'vulnerable' areas in Wigan Borough. A further change was made at this point to the original VLI methodology, with scoring calculated at Super Output Area (SOA) level replacing initial Output Area (OA) level figures. In some instances the relatively small populations contained within Output Areas meant that variances in counts of data sometimes proved too volatile on the impact of the score. By recalculating the scores to SOA level, some of this instability was removed.

10.3 Vulnerable localities in Wigan

The initial scan identified 32 SOAs that scored on average 1.75 or more times the Borough average for the combined VLI score. There were few surprises in these data in terms of the areas identified but what was unexpected was the level of disproportionality highlighted. (One of the 32 SOAs covered Wigan town centre, and this has been excluded from the following summary due to the skew it would give to the figures.) The 31 SOAs analysed covered 15% of the Borough population yet suffered 24% of the Borough's British Crime Survey ten comparator crimes (BCS10 crimes). The crime rate in these areas was 52% higher than the Borough average at 105 per 1000 population. Tellingly, despite a drop in the rate of crime from 2003 to 2004, the 31 SOAs analysed retained a larger proportion of offences, i.e. the gap widened (see Figure 10.2).

Arriving at this conclusion, coupled with the identified disproportion, prompted further work. Before proceeding directly to the 32 SOAs, however, it was important to establish if any other areas suffered disproportion that had not been identified by the VLI. A check of the crime rate amongst the remaining 168 SOAs in Wigan revealed that 13 had a higher rate than the average amongst those initially identified by the VLI. A further nine had a rate similar or just below that of the VLI areas.

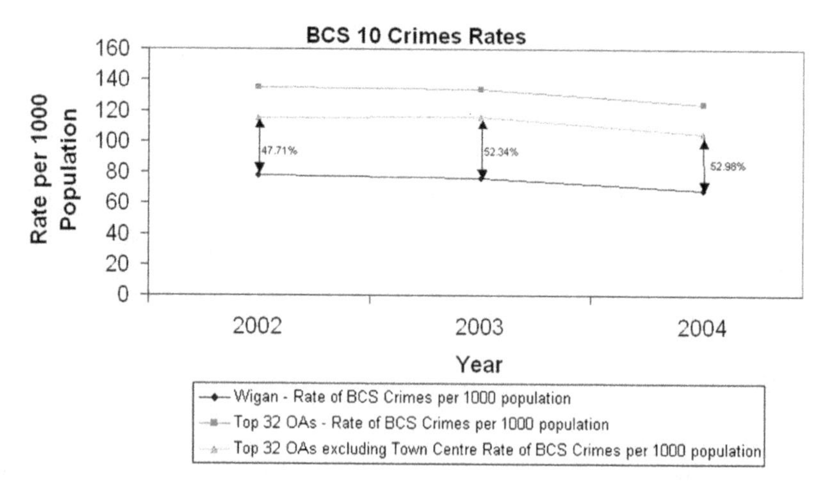

Figure 10.2 Changes in British Crime Survey (BCS) comparator crimes in the 32 top rated Super Output Areas (SOAs), in comparison to the Wigan average.

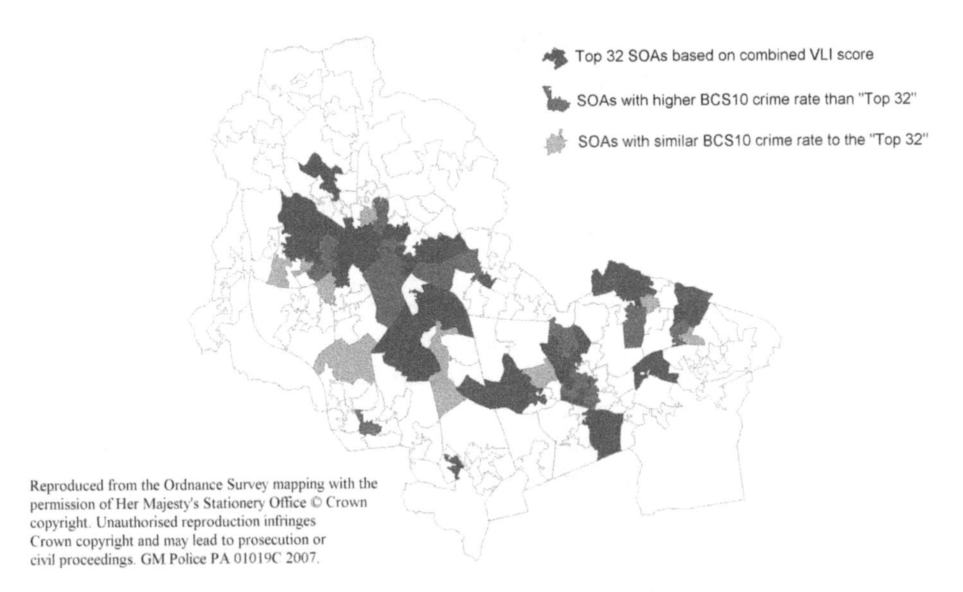

Figure 10.3 Super Output Areas (SOAs) in Wigan identified using the Vulnerable Localities Index (VLI), shown with other SOAs with higher or similar BCS10 crime rates.

(Combined, the 54 SOAs now identified accounted for just over half of all Wigan's BCS10 crimes despite containing just over a quarter of the population.) Interestingly, 21 of these 22 new areas immediately bordered an area identified by the VLI (see Figure 10.3).

This led us to draw some general inferences about these particular areas in Wigan and how they might be interacting with each other. The initial areas identified by the VLI were considered to be the areas suffering more deprivation and the areas that probably housed a larger proportion of the offending population in Wigan. It indicated a need to prioritise people-based interventions in those areas such as targeted drugs outreach, targeted offender work, Progress to Work focus, family support, health issues and housing. Conversely, the areas not identified by the VLI but still with a high crime rate would benefit from a more situational approach to crime prevention such as increased police patrols, alley gates, improved street lighting, improvements to cleaner/greener indicators and improved guardianship.

Clearly, these are generalisations, and much more analysis would be required at a local neighbourhood level to understand the specific nature of offending in each area, but this split between person-focused and situational issues provided a useful starting point. There are also examples in Wigan of this strategic view feeding down to a local tactical level in a similar manner and where the VLI has proved useful in aiding the understanding of offending behaviour.

10.4 Using the Vulnerable Localities Index to help understand offending patterns

In one noteworthy example, an area was identified that was suffering from high levels of acquisitive crime (notably, Burglary in a Dwelling and Theft of Motor Vehicle). There was nothing particularly unusual about the area itself; a series of streets running parallel with one another comprising long rows of terraced housing, similar to a large proportion of other areas across the Borough. The area did not rank highly under the VLI (although it does score above average) and this led us to look at the area in a broader context. Immediately to the north-west of this area is one of the Borough's 'most vulnerable' neighbourhoods (ranked 4th under the VLI) and to the south is the local town centre (see Figure 10.4). Analysis of the crime patterns in the hotspot area revealed a number of interesting facts, but most crucial amongst those facts is that Wednesday was identified as the peak day for offending. Not only that, but the Wednesday peak was a statistically significant one.

The VLI had given us a good understanding of the 'vulnerable' area to the north-west (or our Priority Neighbourhoods, as they later became known). We had a range of indices that gave us a good insight into what contributed to that area's

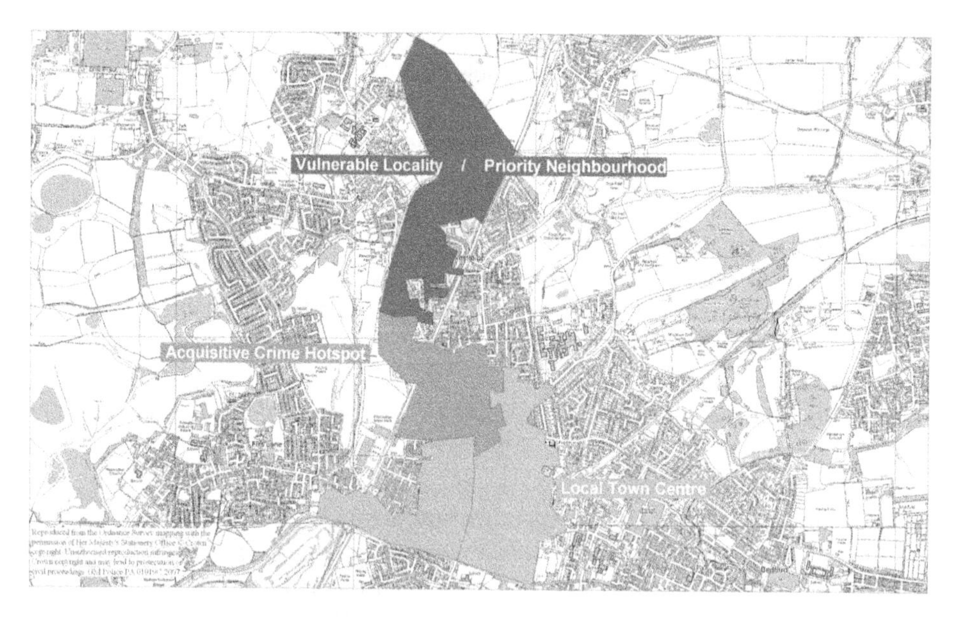

Figure 10.4 Exploring beyond the Vulnerable Localities Index (VLI). This example shows how a vulnerable locality borders other areas with particular crime attributes. In this example it is hypothesised that drug users who lived in the vulnerable locality area were committing acquisitive crimes in an area immediately north of Leigh town centre.

'vulnerability'. Chief amongst these was the prevalence in the neighbourhood of drug users registered with local treatment services (three times the Borough average). Additional research uncovered the fact that only one chemist in the area was able to provide the local methadone scripts to those in treatment. This chemist was based in the town centre and the main pedestrian route from the Priority Neighbourhood to the town centre is via the street between these two areas where crime levels were seen to be extraordinarily high. In addition, the chemist only supplied the prescriptions on one day each week. That day was Wednesday.

In a relatively short space of time, and with only a little additional research, we had been able to develop a robust inference surrounding our crime hotspot, informed principally by our knowledge of the bordering Priority Neighbourhood, i.e. that offenders engaging in drug treatment, or offenders buying methadone from those engaging in treatment, were travelling from the Priority Neighbourhood to the town centre and back, committing acquisitive crime en route in order to raise money to fund drug habits, principally on Wednesdays when methadone could be obtained or purchased to augment their existing drug use.

The strength of this inference allowed us to make very specific interventions across a range of agencies. Notably, we were able to target drugs outreach into the area, spread the distribution of methadone prescription across a range of days so that offenders purchasing methadone were not as certain of being able to buy it at specific times, re-focus the frequency and timeliness of police patrols into the crime hotspot and instigated the initial stages of an alley-gate proposal within each of the terraced rows affected. An ongoing, Borough-wide target-hardening project in conjunction with Victim and Witness Services was also targeted to this area.

The combined effect of this was to immediately reduce burglary and theft of motor vehicle in the area. These reductions have generally been sustained in the two years since the initial analysis took place. It is generally accepted that within this area we have made burglary, in particular, a more difficult crime to commit. In addition, because it was a hotspot for this offence this has impacted at a Borough level as well, contributing to Wigan's continuing burglary reductions.

There is, however, still a note of caution to sound. What does seem to have happened as a result of these actions is that burglary has transferred to within the Priority Neighbourhood and that we have seen a substantial rise in theft from motor vehicle offences across the whole of this area. This suggests that we have made burglary more difficult in the old hotspot area, forcing offenders to commit these crimes closer to home in an environment they are more familiar with and that has received less attention. It also suggests that if they do step outside this 'comfort zone' to offend, then they are more likely to commit an offence that has less likelihood of being caught, such as theft from motor vehicle. These are the next challenges facing the area and serve to highlight that more long-standing issues need to be tackled in Priority Neighbourhoods in conjunction with short-term crime interventions to improve public confidence.

10.5 Developing the Vulnerable Localities Index to support urban regeneration and Neighbourhood Policing

The VLI work in Wigan has progressed since this initial work was conducted in early 2005 and has been used and digested by a number of partner agencies in aiding their understanding of these areas. The rich data source has proved useful to the Local Authority team responsible for identifying areas to benefit from the latest instalment of the Neighbourhood Renewal Fund, and the mirroring of areas identified by other agencies (e.g. Education) has served to add confidence to the way in which each agency is preparing to tackle their most problematic localities.

The introduction of Neighbourhood Policing is also being informed by the VLI (see Centrex, 2006), and it has been a key determinant in agreeing where to focus co-located multi-agency teams in those areas requiring the most intensive partnership support and where Joint Practitioners Groups have been allowed to develop. New approaches, such as Key Individual Networks (KINs), which seek to consult on crime and disorder, have also been directed by the understanding the VLI has given us about our most 'vulnerable' neighbourhoods and will hopefully prove to be an integral tool in the move to partnership neighbourhood management.

10.6 Acknowledgement

The author is grateful to Paul Whitemoss, Research Co-ordinator and Karen Pennington, Research Analyst from Wigan Metropolitan Borough Council for their help in preparing the material for this case study.

10.7 References

Centrex (2006) *Briefing paper on Neighbourhood Policing and the National Intelligence Model.* Wyboston: Centrex.

Chainey, S. P. (2004) Using geographic information to support the police response to community cohesion. Paper presented at the *Association for Geographic Information Conference*, 12–14 October, London.

11 Reducing re-offending in local communities: geographical information system based strategic analysis of Greater Manchester's offenders

David Ottiwell

11.1 Context and introduction

This is a case study that documents how an analytical team based at Greater Manchester Police headquarters worked within a nationally recognised Partnership Business Model to develop new techniques in the analysis of information about offenders. The work outlined was begun at a countywide level in spring 2005, developed at a borough level throughout 2006, and continues to be relevant to the agenda today.

Looking retrospectively, the analysis was prompted not by a single commissioning meeting or agreed terms of reference, but rather by the constellation of a number of factors enabling and demanding a GIS-based strategic analysis of offenders. These included:

- the national Reducing Re-offending agenda;
- 'New Localism' and the expanding discipline of 'area profiling';
- partnership data availability and GIS specialism in Greater Manchester.

Crime Mapping Case Studies: Practice and Research Edited by Spencer Chainey and Lisa Tompson

A particular influence for the work was the Local Government Association report *Going Straight*, published in February 2005. This scoped out the expanded future role of local government in delivering reduced re-offending, by highlighting the well-rehearsed failings of the criminal justice sector alone to rehabilitate offenders and address the underlying causes of their problem behaviour. A key message from the report was set out as follows:

Local authorities should consider what opportunities they have to prevent offending and reduce re-offending by targeting their own services – notably housing (including benefits), education, social services, employment (including as employers in their own right), community development/regeneration, leisure and, of course, community safety – towards preventative interventions or dealing with specific offender needs (LGA, 2005, p.4).

In the context of prior knowledge of the national reducing re-offending framework, this sets delivery against the seven rehabilitation pathways (accommodation; education, training and employment; mental and physical health; drugs and alcohol; finance benefits and debt; children and families; attitudes, thinking and behaviour) in the context of getting services right at a community level. In essence, it stresses that the commissioning of key community services should be evidence-based and in accordance with identified local need. *Going Straight* (LGA, 2005), however, does not elaborate on the distinct information sources that might be used to conduct local needs assessments.

The Greater Manchester analytical team decided that information shared under partnership protocols across the conurbation could be used to understand risk factors relating to re-offending in a spatial manner, rather than on a person-by-person basis. In particular, established offender risk assessment tools used by criminal justice bodies such as the National Probation Service and by Youth Offending Teams (YOTs) were seen as the key datasets for this work. Put to a different purpose, a comprehensive risk assessment of volume offenders could be geographically enabled to give a strategic overview of where crime drivers concentrate and where partnership resources are most needed.

11.2 Implementation, testing and analysis

11.2.1 Analytical methodology

The analysis was developed in relation to three main areas.

- Risk profiling of Prolific and other Priority Offender (PPO) cohorts, and supplementary to this, of geographically distinct PPO 'catchments' using home address information.

- Identifying the spatial distribution of known offender residences, for any bespoke offender 'type' or 'grouping' – e.g. problematic drug users.
- Producing partnership neighbourhood problem profiles that showed 'Safer Neighbourhoods' in terms of re-offending risk factors applicable to resident young people.

The exact methodology in relation to each of these areas has varied slightly, but each has relied on two key precursors, as follows:

- access to Probation Service and YOT data;
- geocoding of offender data.

11.2.2 Access to Probation Service and Youth Offending Team data

The supply of detailed information from both the National Probation Service for Greater Manchester, and from local Youth Offending Teams, has undoubtedly been made possible by the development of Greater Manchester Against Crime (GMAC). This has applied the UK Police National Intelligence Model (NIM) in a partnership context and, critically, secured the agreement of key service providers to contribute data to a common data repository held on a secure central server. These data can be accessed remotely across the ten boroughs of Greater Manchester by any of the dedicated Strategic Analysts who sit within local authority Community Safety teams, or at the centre looking at countywide issues.

In relation to the Probation Service, GMAC contracts out certain data sharing responsibilities to the Association of Greater Manchester Authorities (AGMA). This body has a responsibility to take a sensitive data extract from the Probation Service on a quarterly basis, create a semi-depersonalised subset of data in accordance with an agreed GMAC Data Sharing Schedule, and add this to the central data hub. At the same time, AGMA retains a copy of the full, sensitive data that is not shared as a matter of course, but can be released with agreement for the purposes of specific pieces of work. The data provided by the Probation Service is a full slice of electronic Offender Assessment System data (OASys). The OASys is a risk assessment and sentence planning tool, developed jointly by the Probation and Prison services, to assist practitioners in identifying and classifying offending related needs. In its entirety, the dataset supplied covers twelve sections around risk of re-offending, a substantial list of variables in relation to risk of harm and vulnerability, and various other useful markers. In all, the full dataset incorporates around 600 data items; the reduced dataset covers around 100 items.

Attempts are ongoing presently within GMAC to secure a common dataset for all boroughs in Greater Manchester that reflects the youth offending picture. While this continues, however, work has begun through local arrangements, whereby some analysts have been granted access, directly or indirectly, to the Youth Offending Team's integrated case management and risk assessment system. The Youth Offending Information System (YOIS) is a software solution provided to YOTs around the country. Work in this case study was enabled by the agreement (formalised again in a detailed data sharing schedule) that the Crime and Disorder Reduction Partnership (CDRP) analytical team could access ASSET risk assessment data, much in keeping with its adult equivalent, OASys, for the purposes of research and youth crime prevention work. ASSET risk assessments cover a similar number of theme areas as OASys assessments (including living arrangements, family and relationships, education, substance misuse), but tend to have a simpler mechanism for calculating risk 'scores'.

11.2.3 Geocoding of offender data

The GIS-based analysis used information held about offender home addresses to anchor the information about risk factors and offending propensities in a geographical context. This has been achieved by exporting the full home address postcodes of offenders (i.e. M16_0RE) within the datasets into MapInfo GIS and adding x and y coordinates to these postcodes by reference to a master file called Code-Point. Code-Point is an Ordnance Survey gazetteer product providing a precise geographical reference for each postcode unit in Great Britain. It is available to Police Forces and Local Authorities under the terms of the local Mapping Services Agreement. As the work was strategic rather than tactical, it was felt that mapping to a unit postcode level of accuracy was more than adequate, given that there were typically only 12–15 unique households in a postcode. The analysis outlined in this case study found that postcodes could usually be matched with more than 90% accuracy. Difficulties that arose were normally in connection with offenders registered as currently of 'no fixed abode', or where offenders were currently serving custodial sentences or on remand, with no usual residence in the community provided.

11.2.4 Profiling prolific and other priority offenders (PPOs)

The first achievement of our GIS-based analysis of PPOs was to represent the spatial distribution of the PPO home addresses in a far more meaningful way than was previously the case. Instead of working with a choropleth map showing a count of PPOs by the home policing subdivision, kernel density estimation (KDE) was used to show the concentration of residences irrespective of administrative geography. This made the information accessible to crime and disorder partners not familiar with policing geography, and gave a granularity which identified several key (if not unsurprising) deprived urban areas and neighbourhoods peripheral to Manchester city centre, and in outlying parts of Manchester and Salford.

Gaining a more precise geographical understanding of where this key cohort of offenders were living was also useful in demonstrating their collective disproportionate residence in high priority communities already identified as critical to the common crime and disorder aims of the Partnership. Building on the development of the Vulnerable Localities Index (VLI) (Chainey, 2004) as the principal tool used by Greater Manchester's partners to identify localities most at risk of community breakdown and fragmentation, the analysis showed that over a fifth of currently or previously targeted PPOs were residents in only 4% of the Partnership's most vulnerable localities. It suggested that if the Partnership agreed to concentrate most resources in its top 16% high crime–high disadvantage areas, it would probably be investing in those localities where over half of all offenders on the PPO scheme since September 2004 were resident.

Although the partner data provided on the GMAC data hub is ordinarily stripped of all unique identifiers, the key to risk profiling the PPO cohort (defined primarily in policing systems) was to match the PPO records against the Probation data using a common reference number held in both datasets – the Police National Computer Number (PNC-ID). The data sharing schedule in place for sharing the data reflected this and GMAC analysts have been trained in appropriate data usage. Once the two datasets were merged, analysis could be undertaken on geographical, postcode level risk assessments. The data for the entire PPO cohort showed that 49% had accommodation issues serious enough to contribute to the heightened risk of re-offending. This was 85% in relation to education, training and employment. However, this information was not particularly helpful in defining the problem and scoping possible responses at the local level. What the geographical data enabled the analyst to do was take several chosen offender clusters (e.g. compare PPOs in Wythenshawe, with PPOs in Moss Side, with PPOs in Gorton) to determine in which areas these problems were most sharply exhibited. This could then start to paint a picture in terms of offender needs assessments, for use by local PPO teams looking to deliver against the PPO Strategy.

11.2.5 Identifying local drugs markets

Moving away from a PPO focus, it was clear to the analytical team that the Probation risk assessment information had the potential to profile entire offender populations across Greater Manchester in accordance with the needs of the Partnership. This became of direct relevance to work commissioned by senior police officers within

Greater Manchester Police to conduct the first comprehensive, countywide profile of class A drugs use – a profile of criminality associated with drug use, dealer networks, drugs markets, criminal justice intervention (mandatory testing, arrest referral), treatment and need.

Analysis was conducted to review all adult offenders assessed by the Probation Service over the previous twelve months. A subset of data was taken that represented any offender assessed as being either a current or previous user of illegal substances. The data could then be mined and cross-tabulated according to a wealth of further information detailing trends in drug use. Maps of offender home address clusters were produced using KDE. The KDE outputs were sensitive enough to show the subtle differences between various mapping outputs that were loaded as images into a rolling Microsoft PowerPoint file: male drug user versus female user hotspots, heroin user versus crack cocaine user hotspots, injecting population hotspots, poly-drug user hotspots, and high risk drug user hotspots (risk of reconviction and/or risk of serious harm). Collectively, the analysis was able to profile the known drugs markets throughout the conurbation with a resolution not previously available. Sat alongside a strategic overview of acquisitive crime hotspots and vulnerable communities, the hypothesised dependencies between drug use, criminality and victimisation could be drawn out and summarised.

11.2.6 Risk factor neighbourhood profiles

The third area where offender data augmented strategic understanding relates to the issue of neighbourhood profiling. The use of YOT risk assessment data in a geographical context was an important means of showing that neighbourhood profiles considering police crime and incident data were insufficient to reflect the wider responsibilities of crime and disorder partners. Association of Chief Police Officer guidance on the issue of neighbourhood policing has tended to focus upon applying a fairly narrow NIM approach to neighbourhood profiles that presents crime trends alongside conventional (graded) intelligence sources, possibly in conjunction with geodemographic data from the 2001 census and other commercially available data sources. The GMAC analysis already produced using Probation Service data raised the possibility of profiling small geographical areas according to the current risk profile of associated offender cohorts.

Analysis was undertaken locally in one CDRP in Greater Manchester. This area was chosen because it was particularly keen to understand how Partnership neighbourhood policing delivery could align more effectively with the work of professionals engaged with young people at risk of entering the criminal justice system. A three-year extract of YOT ASSET risk assessments was mapped to the relevant home address postcode of the young person in question. These data (composed of almost 1000 offenders) was given a geographical profile by assigning each young person to one of thirty-three 'Safer Neighbourhood' areas.

Not all neighbourhood areas were host to young offenders, so the decision was taken to use the YOT data to profile six neighbourhoods already identified as priority areas based on recorded crime and anti-social behaviour. These areas were also those with the highest volume of associated young offenders in the YOT data. For each neighbourhood (each with at least 75 offenders), calculations were undertaken to determine the proportion of resident young offenders who had been assessed as higher risk in relation to twelve different risk factors. So, for example, it could be determined that almost 70% of young offenders in one particular neighbourhood had issues around the 'thinking and behaviour' factor that were strongly, or very strongly associated with their risk of re-offending – well above the borough average or for any other area. Two characteristics that were highlighted as high priority issues were 'family and personal relationships'. The profiling exercise also overturned some preconceptions around substance misuse, where a neighbourhood on the (generally) more affluent side of the borough matched more disadvantaged, inner-urban neighbourhoods in terms of associated re-offending risks.

11.3 Results, outcomes and issues

Outcomes associated with the analysis have been both strategic and operational at the neighbourhood level. The work around class A drug use has fed into Greater Manchester's drugs strategies, helping Partners to understand drugs markets and therefore to balance police disruption tactics with the need for locality based drug treatment interventions. In Wigan, local work using Probation data has examined high-risk problematic drug users' home addresses in the context of their daily routines and journey to crime in the town centre, leading to a problem solving debate around the future location of prescribing services. In Bolton, the analysis of youth offending risk factors has led to the targeted commissioning of services to support 'positive contributions' work with young people. The work has been used as an evidence base underpinning service bids for Neighbourhood Renewal Fund monies targeted in priority neighbourhoods. It continues more generally to inform the wider debate about agencies pooling resources and commissioning services jointly in identified areas of need.

Issues do remain though in wholesale access to data systems, particularly in relation to youth offending and YOIS functions, both as a case management tool and as a repository for risk assessments. Therefore data regarding the prevention and detection of crime sits alongside very sensitive information relevant to sentence planning and the private lives of young people. The next challenges for the analyses are to consider the spatial analysis of offender data in conjunction with a spatial understanding of existing resources and partnership assets.

11.4 References

Chainey, S. P. (2004) Using geographic information to support the police response to community cohesion. Paper presented at the *Association for Geographic Information Conference*, 12–14 October, London.

LGA (2005) *Going Straight – Reducing Reoffending in Local Communities* (online). London:. Local Government Association. http://www.lga.gov.uk/Documents/Publication/goingstraightfinal.pdf

Part IV Integrating visual audits and survey data into crime mapping

12 Community Safety Mapping Online System: mapping reassurance using survey data

Steven Rose

12.1 Introduction

In recent times a shift in government policy has necessitated the acceptance by Police Forces and Partner Agencies that public reassurance is a key delivery objective in achieving community safety (Home Office; 2004, 2005). Reductions in crime are almost meaningless if the public we serve do not 'feel the difference' and remain fearful. This is backed up by a need to understand and tackle the growing gap between falling levels of crime and increasing concern about crime and disorder. At the heart of reassurance and perception is the theory of 'signal events' (Innes and Fielding, 2002) – those types of crime, disorder and other events that matter to the public and those which can have a negative impact on people's perception of safety and police performance. Signal crimes have a disproportionate impact on people's feelings of security. Examples being issues such as graffiti, vandalism, abandoned vehicles, litter and youths hanging around the streets, all of which can generate a lot of anxiety. Results from a survey commissioned by West Midlands Police in Birmingham, UK, the 'Feeling the Difference Survey', showed that 36% of respondents list 'troublesome teenagers' as the greatest crime and disorder issue in their neighbourhood, closely followed by 29% who list vandalism and graffiti.

This case study will detail an analytical research project that draws on two distinct surveys undertaken in Birmingham, based on perception and environmental quality, and how through the innovative use of COSMOS (Community Safety Mapping

Crime Mapping Case Studies: Practice and Research Edited by Spencer Chainey and Lisa Tompson

Online System) a clearer picture of the landscape of reassurance is being mapped out. Following on from this, the picture generated is assisting in the identification of the key drivers of fear of crime, providing a practical means to effect a change in public reassurance at a neighbourhood scale.

12.2 Community Safety Mapping Online System (COSMOS)

Community Safety Mapping Online System is a website that has been created for the Birmingham Community Safety Partnership and is designed as a central point of contact for agencies in Birmingham and the West Midlands. The system provides access to multi-agency data through standard reports and easy-to-use mapping and analysis tools. Data are reported in standard agency geographies, allowing data and information to be compared and analysed together while maintaining maximum flexibility. The website is a tool to aid true partnership working and information sharing, and with its intuitive interface is not the preserve of only the PC literate. Through consultation with the Crime and Disorder Reduction Partnership (CDRP) and its representatives, COSMOS is designed not as an analytical tool but with the end-user in mind. The aim is for key partnership information to be available to decision makers at the click of no more than a few buttons. Also key to the success of COSMOS is the team of partnership analysts that the Birmingham Community Safety Partnership employs within its Information and Intelligence Team. The analytical resource works in tandem with the scanning and management information that COSMOS provides. Where a need for detailed analysis or problem solving is identified through COSMOS, the Information and Intelligence Team have the required skills to be able to support such demands. The model of COSMOS combined with the Information and Intelligence Team gives the Partnership a holistic approach in providing a sure evidence base to assist in the delivery of its strategic aims (for more details on COSMOS see Chainey and Smith, 2006, or visit http://www.jdi.ucl.ac.uk/ho_sharing_network/sharing_systems/index.php).

12.3 Measuring reassurance

To measure reassurance, an extensive household survey that assesses some key factors was commissioned by West Midlands Police. This included:

- the public's perception and fear of crime;
- community cohesion;
- the service provided by law and enforcement agencies.

This 'Feeling the Difference Survey' has now been through 10 waves, each wave consisting of over 5000 respondents, resulting in more than 50,000 responses to date. The waved approach allows an assessment of perception over time. The need to be able to monitor progress is vital in order to be able to perform results analysis and determine shifts in reassurance. The survey has been conducted to allow comparison of 21 different Operational Command Units (OCUs) across West Midlands Police Force. Crucially, however, there was no means to be able to analyse the results at the neighbourhood level. This is often a limitation to survey results that is all too apparent to the practitioner working at the local level. With the introduction of neighbourhood policing and neighbourhood management, the buy-in of these precious resources is paramount. As such a clear local understanding of reassurance provides practical added value to the survey.

To overcome the problem of the survey being reportable at OCU level and not for smaller areas and to really bring to life the potential of the data, COSMOS was used. To be able to map at the local level, survey results were extrapolated using a geodemographic tool ('ACORN' from CACI solutions) that can be used to interpolate respondents' propensity to answer in a given way. Each postcode is assigned an ACORN type classification. This classification is built from a wide range of socio-demographic lifestyle data that segment the population into groups with particular similar traits (for more details see http://www.caci.co.uk/acorn/). An example being ACORN type 'Wealthy Executives' who have a higher propensity to be directors, have large homes and go on winter skiing breaks. It also turns out that by matching your respondents' survey results to their ACORN type, clear behavioural differences become apparent. Returning to the 'Wealthy Executives', as a group they are over three times as likely as the average person in Birmingham to agree that they live in a harmonious area. This makes it possible to perform analyses of perceptions at geographies as small as a postcode (15 households). By looking at the mix of postcode types in a neighbourhood, a propensity to respond to the survey in a given way can be calculated. We can now provide a clear picture of the subtle nuances of perception at the neighbourhood level.

A practical example of the use of this information in Birmingham is as follows. A particular neighbourhood was highlighted by the 'Feeling the Difference' results, through COSMOS, as being of a lower propensity to agree that people from different backgrounds live together harmoniously within the neighbourhood (Figure 12.1). Overlaying this information with other intelligence and data sources showed that the area had significant support for nationalist political parties as well as being a cluster area for known offenders and victims of hate crime (primarily related to a race or religious motive). Clearly the survey based on qualitative information provided through COSMOS is combining with the hard quantitative data to sure up the evidence to prioritise this neighbourhood in terms of community cohesion.

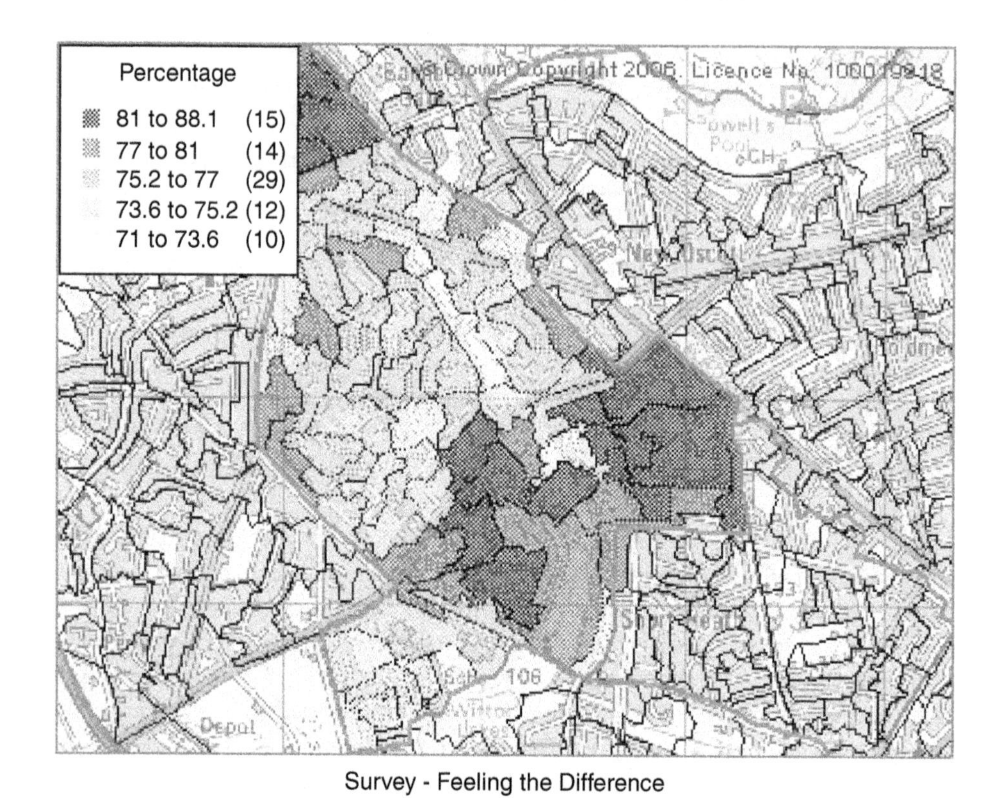

Figure 12.1 Mapped results from Birmingham's 'Feeling the Difference' survey showing different neighbourhood levels of residents saying they live 'harmoniously'.

12.4 Environmental visual audit (EVA)

To measure environmental quality, Birmingham City Council commissioned ENCAMS (better known as the 'Keep Britain Tidy' group) to complete a comprehensive environmental visual audit of the city. It was recognised that the survey results were potentially of greater value if it was possible to map these results. Similar issues of mapping at a neighbourhood level as with the public perception survey were preventing the full buy-in and utilisation of this valuable intelligence. ENCAMS approached Birmingham Community Safety Partnership to devise a means to spatially enable the survey results using COSMOS.

The EVA comprised a comprehensive survey of visual indicators of degradation and signal events in Birmingham. ENCAMS' unique Local Environmental

Quality Survey (LEQS) Standard Quality Interval (SQI) approach to measuring environmental quality shows a clear cognitive step between good levels of environmental quality to poor. Data were recorded along 50 m transects across the city ensuring a representative sample of distinct land-use types. ENCAMS research has shown that land use is a vital part of environmental monitoring as it acts as a good predictor of environmental quality. Where the presence of signal events and/or environmental degradation is discovered within a particular land use, areas nearby of a similar land use will have a propensity to suffer similarly with environmental issues. In other words the environmental quality recorded from the survey transect of a high density residential street is likely to be replicated for the high density residential streets in close proximity. The results provide a snap shot of environmental quality and can be used as a tool to prioritise reassurance policing and a multi-agency targeted response.

A key desire for the Partnership, to assist in strategic decision-making and resource deployment, was the ability to map the results and analyse environmental quality at the local neighbourhood level. Extrapolation of the EVA results was possible using land use as a predictor of environmental quality. This means the results can be mapped to show the prevalence of signal crimes and environmental degradation within neighbourhood communities of Birmingham. This clearly avoids the impracticality and significant cost of completing a 100% coverage environmental audit of the whole city. Environmental quality can now be mapped and interrogated using mapping tools such as a gauge chart which gives a visually appealing illustration of environmental quality across a suite of indicators for an area at the click of a button as the user explores neighbourhoods (Figure 12.2).

12.5 Practical use of results

The two surveys combine to become a critical data source providing a solid evidence base to deliver the objective of improving public reassurance. A key example of the use of the survey data is in the Strategic Assessment of community safety in Birmingham. The Strategic Assessment draws on relevant information to assess the critical themes and neighbourhoods where community safety issues are disproportionate. Previously it has been difficult to gain a holistic view of both recorded crime and disorder events in conjunction with the softer elements of fear of crime and signal events such as environmental degradation. With the innovation through COSMOS it has been possible to analyse the interaction for both recorded incident hotspots and community reassurance. In Birmingham we were able to draw up a set of priority neighbourhoods that have cross-cutting community safety priorities including an understanding of public reassurance at the neighbourhood level. As

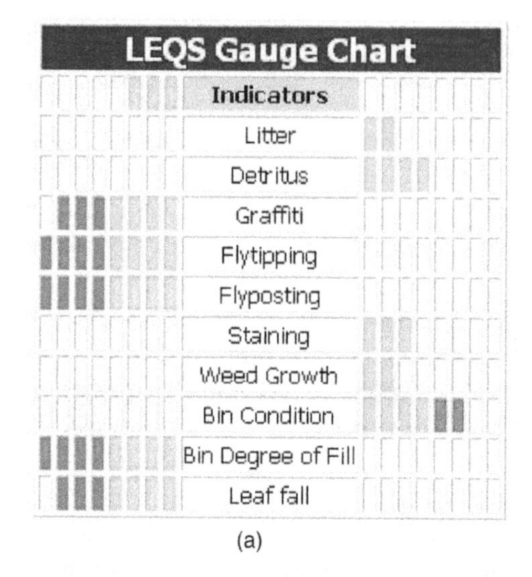

(a)

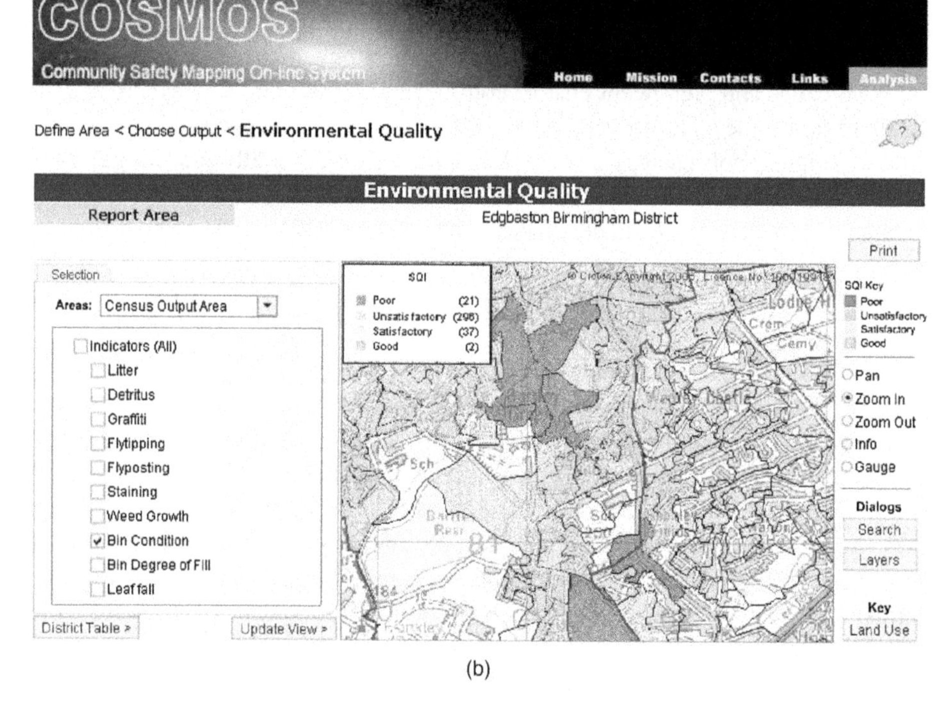

(b)

Figure 12.2 Environmental quality visual audit results shown as (a) a gauge chart and (b) as a map.

such there is an improved evidence base to assist in the formulation of strategic policy and the performance management of Birmingham's community safety partnership strategy for 2005–2008. This then feeds into informing a local tasking and co-ordination system within the Birmingham Community Safety Partnership.

The Birmingham Community Safety Partnership Information and Intelligence Team use a composite index of community safety to assess key locations and themes within the city of Birmingham. Both the perception and environmental survey data mapped to local areas have added greatly to the understanding of the mechanisms behind perceptions of fear and the practical responses most appropriate to effect a change in public perception. A study completed as part of the Strategic Assessment process aimed at finding which community safety issues most closely correlated to public perceptions of fear by community. The feeling the difference survey included a question specifically asking 'how fearful are you of becoming a victim of crime?'. A host of variables were compared against the fear results including troublesome teens, vandalism and graffiti, anti-social behaviour, criminal damage, street crime and many others. The research identified some interesting correlations. For example the community safety indicator that most closely corresponded to people's fear of being a victim of crime was recorded criminal damage to dwelling (see Figure 12.3). Communities that had the highest fear levels suffered the highest levels of criminal damage to dwelling.

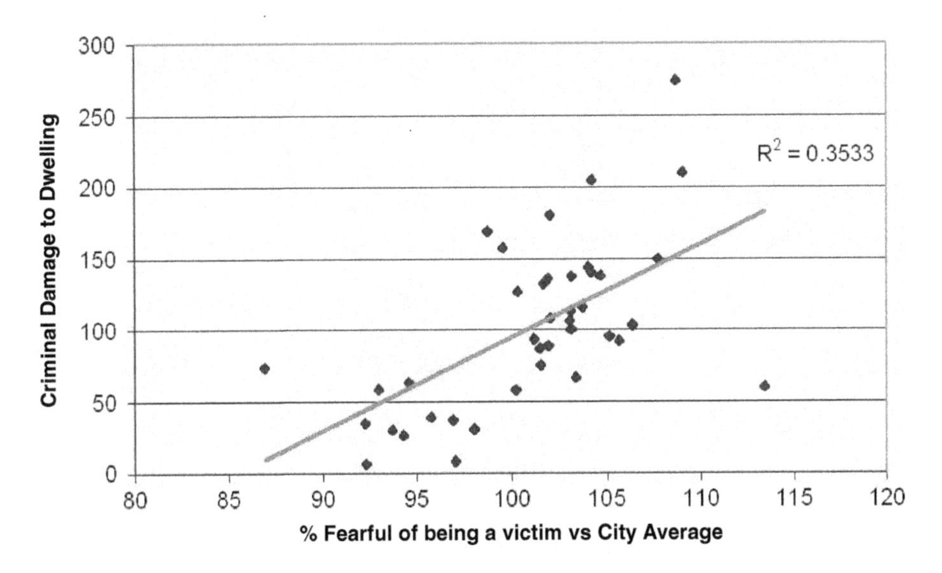

Figure 12.3 Correlation analysis between fear of being a victim of crime against criminal damage for neighbourhoods in Birmingham.

The recorded number of anti-social behaviour incidents was also a variable that closely correlated to peoples' level of fear. Interestingly some myths or red herrings were highlighted. For example the level of litter in a community did not correlate with the level of fear. This does not mean that litter is not important, but was not as useful a mechanism to prioritise activity or gain maximum leverage in effecting people's fear of crime. In Birmingham the survey shows that the greatest problem that communities perceive is that of troublesome teenagers. This is consistently listed as the greatest concern for people. The extent to which a community perceive teens as the issue, however, does not correlate with their fear of crime. As such, a heightened perception of troublesome teenagers does not automatically translate into fear. All communities perceive teens as the problem, so by using this as a variable to prioritise community safety issues does not make it possible to differentiate places of greater need or hope to most effectively alter perceptions of fear. The key is in what the teenagers are actually doing, which comes back to the issues of criminal damage to dwelling and anti-social behaviour. Birmingham now has a clearer understanding of which are the stronger signals of fear in the community and can prioritise actions to best improve feeling safe within local neighbourhoods.

12.6 Evaluation and next steps

The ongoing results from the feeling the difference surveys have shown marked reductions in the fear of crime within Birmingham. Table 12.1 shows the significant improvements made on the three key perception based performance indicators in Birmingham, both in terms of feelings of safety and people from different backgrounds getting on well together in their communities.

Table 12.1 'Feeling the Difference' survey results for Birmingham, West Midlands (Source: West Midlands Police 'Feeling the Difference' Survey)

Survey Question	Waves 1–4 (2004–05) Percentage that answered 'safe'	Waves 8–11 (2006) Percentage that answered 'safe'	Change (%)
How safe or unsafe do you feel when outside in your neighbourhood during the day?	89.4	94.0	+4.6
How safe or unsafe do you feel when outside in your neighbourhood after dark?	55.2	61.0	+5.8
To what extent do you agree that your neighbourhood is a place where people from different backgrounds and communities can live together harmoniously?	81.3	92.0	+10.7

The need to better understand reassurance in the community demands an ever increasing interaction with the community. Both the 'Feeling the Difference' and environmental surveying will continue. An exciting commitment to the environmental audit through Birmingham City Council is the development of an ongoing survey methodology. Environmental surveyors have been employed to continually survey the city by carrying hand-held mapping devices, which allow the accurate recording of environmental indicators and signal crime information. The data collected will be fed into COSMOS, to allow progress monitoring and change mapping. The Birmingham Community Safety Partnership will then be able to evaluate its progress and best coordinate the strategic planning of reassurance based neighbourhood management. West Midlands Police, in their desire to improve public reassurance and interact better with policing neighbourhoods, have also developed a public access website called www.myneighbourhood.info using the same engine and architecture as COSMOS (Figure 12.4). The site maps crime within local neighbourhoods, shows crime trends and provides a 'toolkit' of bespoke local links to community initiatives as well as national providers such as Victim Support. The toolkit also offers interactive guidance on crime prevention. The website allows users to complete online

Figure 12.4 The Birmingham Community Safety Partnership's MyNeighbourhood website, providing local crime statistics to the public.

polls and surveys, allowing a dynamic and continual assessment of reassurance and perceptions of community safety related issues. All of the intelligence gathered will continue to feed back into the analytical problem solving processes of the police and the community safety partnership in Birmingham.

12.7 References

Chainey, S. P. and Smith, C. (2006) *Review of GIS-based Information Sharing Systems*. London: Home Office.

Home Office (2004) *Confident Communities in a Secure Britain*. The Home Office Strategic Plan 2004-08 (online). London: The Stationery Office (CM6287). http://www.archive2.official-documents.co.uk/document/cm62/6287/6287.pdf [Accessed 14 June 2007].

Home Office (2005) *National Policing Plan 2005–2008* (online). London: The Stationery Office. http://police.homeoffice.gov.uk/news-and-publications/publication/national-policing-plan/national_policing_plan.pdf [Accessed 14 June 2007].

Innes, M. and Fielding, N. (2002) From community to communicative policing: 'signal crimes' and the problem of public reassurance. *Sociological Research Online* **7**(2). http://socresonline.org.uk/7/2.

13 Mapping the fear of crime – a micro-approach in Merton, London

Chris Williams

13.1 Introduction

Reducing the fear of crime has become a cornerstone of British social policy in recent years. The new National Community Safety Plan has the reduction of the fear of crime at the heart of its message: 'The Government's key priorities for 2006–09 are to . . . continue to develop, evaluate and disseminate good practice on reducing crime and the fear of crime' (Home Office, 2005).

The fear of crime can be considered as 'an anticipation of victimisation, rather than fear of an actual victimisation' (John Howard Society of Alberta, 1999). With a recorded reduction in crime rates in England and Wales of 35% since 1997 (Blears, 2005), one might expect to have observed a corresponding fall in the fear of crime. However, this appears not to be the case.

In Merton, South London, residents' concerns about crime actually rose 6% between 2004 and 2005 according to the Merton Annual Resident's Survey (TNS Social, 2005). Although this could be due to many factors – media images, a rise in environmental and so-called 'signal' crimes (Innes and Fielding, 2002) – one cause that can be discounted is an actual rise in reported crime; in Merton this fell by 7.8% over the same time period (Wood, 2005). As Innes and Ditton (2005) state: 'There has never been much relationship between levels of crime and levels of fear, and . . . reductions in crime have rarely led to reductions of fear of it'.

This anomaly is known as the 'Reassurance Gap', and is one which government and police services are determined to close. A population believing that crime is

Crime Mapping Case Studies: Practice and Research Edited by Spencer Chainey and Lisa Tompson

increasingly prevalent is likely to think unfavourably of its elected representatives and the local police service with the expressed purpose of ensuring their safety.

13.2 Process

Within this context, in Merton a partnership between the local Basic Command Unit (BCU) of the Metropolitan Police and the Local Authority's Crime and Disorder Reduction Partnership (CDRP) sought to tackle the issue of fear of crime in one particular area of Abbey ward, where traditionally burglary had been common. This particular group of eight streets is sandwiched between a run-down area of council housing containing three tower blocks and an area of very expensive privately owned terraced housing.

The area had been selected for use in a trial of a property-marking scheme in order to attempt to reduce the number of burglaries within these eight streets. As the police beat officers would be required to visit each occupied property on each street, it was realised that a written questionnaire, completed by the residents when the police were in attendance, could be carried out simultaneously in order to ascertain the fear of crime accurately in each household. Consequently, a 98% response rate was obtained from 309 households. There is a potential confounding bias with this approach; the very questioning of an individual about crime can negatively affect their feelings about the subject; the so-called 'Hawthorne Effect' (i.e. interviewer bias). For these purposes, however, internal rather than external validity is the key. In summation, we were seeking to complete a 'Fear of Crime Census' within parts of the Abbey ward.

In order to establish some rationale above and beyond the quantitative approach, we were able to examine the dialogue replies that were captured on the questionnaires. The data that was extracted from these responses – though largely anecdotal – has in turn been assisted by the interpretation of results that are illustrated from the mapping process that has been applied. The process is explained in the following section in more detail but essentially requires the questionnaires to be analysed, which then feeds into the results being mapped, the maps to be explored and interpreted, helping to then direct action by the relevant policing and crime reduction agencies.

13.3 Methodology

A questionnaire was distributed to each of the 309 households in the selected eight streets in Abbey ward, and asked, 'How do you feel about . . .' the topics listed in Table 13.1, which are historically the main concerns outlined to us by residents in public meetings and surveys. A 1 to 4 scale was used to ascertain people's levels of concern about each crime type. Each response was weighted, so that a major

Table 13.1 The questionnaire used in Merton to ask residents how they felt about crime and anti-social behaviour issues

"How do you feel about. . . ."	Major problem	Minor problem	No problem	Don't Know
Crime	1	2	3	4
Youth anti-social behaviour	1	2	3	4
Burglary	1	2	3	4
Robbery				
Mugging or pick pocketing	1	2	3	4
Car crime				
Stealing cars or from cars	1	2	3	4
Drugs	1	2	3	4
Alcohol-related problems	1	2	3	4
Hate crimes				
Crime motivated by a hatred of difference, like the colour of someone's skin, sexual preference, religion or disability	1	2	3	4
Traffic problems				
Speeding, parking, dangerous driving	1	2	3	4
Vandalism and graffiti	1	2	3	4

problem was worth 3 points, a minor problem 2, and no problem 1, from which a mean value was then calculated.

As the mean values produced in each instance were very close to both the mode and median, we could state that our data were normally distributed and we could use the mean values on an interval scale. Previous analyses have shown little difference between the mean and the median, and for illustrative purposes the mean has been used here to show levels of fear of crime.

The results of the survey were collated on a street-by-street basis, and then mapped using the thematic-mapping tool in MapInfo, before being illustrated in the format shown in Figures 13.1–13.3. In each map, levels of fear were articulated by darkening graduations; so the darker the colour, the higher the level of fear. The graduations themselves were those automatically designated by the MapInfo (i.e. the equal count range method), generally dividing into four blocks of two streets each. Below are three examples of the results.

13.4 Results

13.4.1 All crime

Figure 13.1 shows the overall fear of crime in the eight streets selected for the pilot. The streets with the highest fear of crime were Haydon's Road, Wandle Bank and Grove Road.

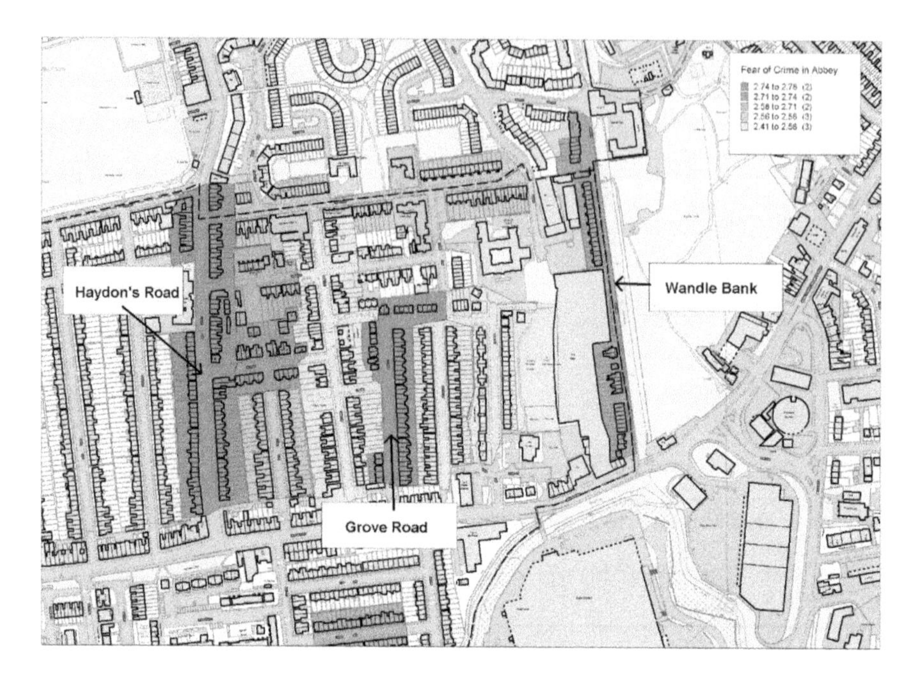

Figure 13.1 Fear of all crime in the Merton study area, London.

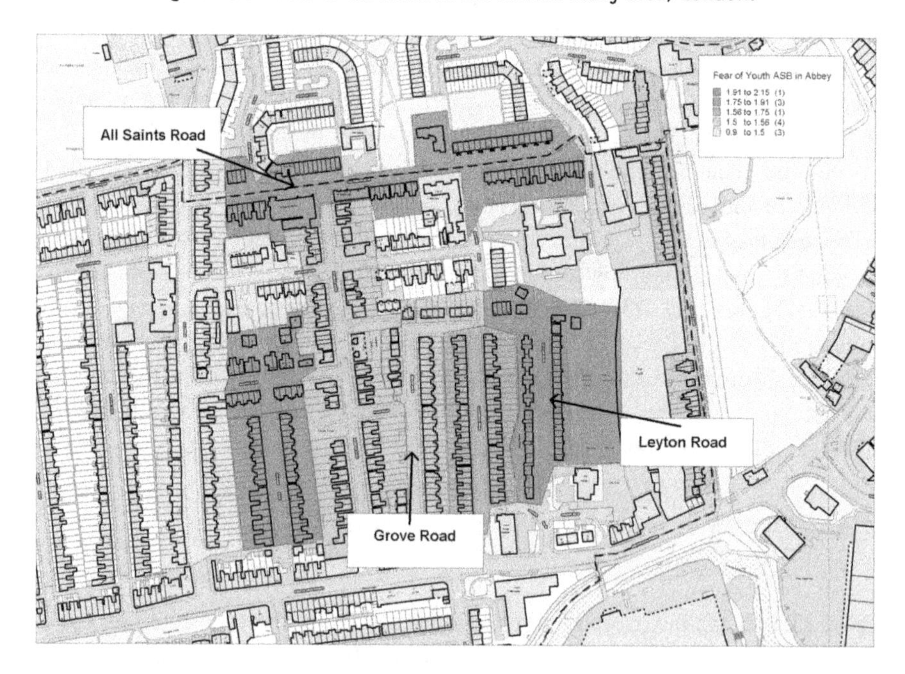

Figure 13.2 Fear of youth anti-social behaviour in the Merton study area, London.

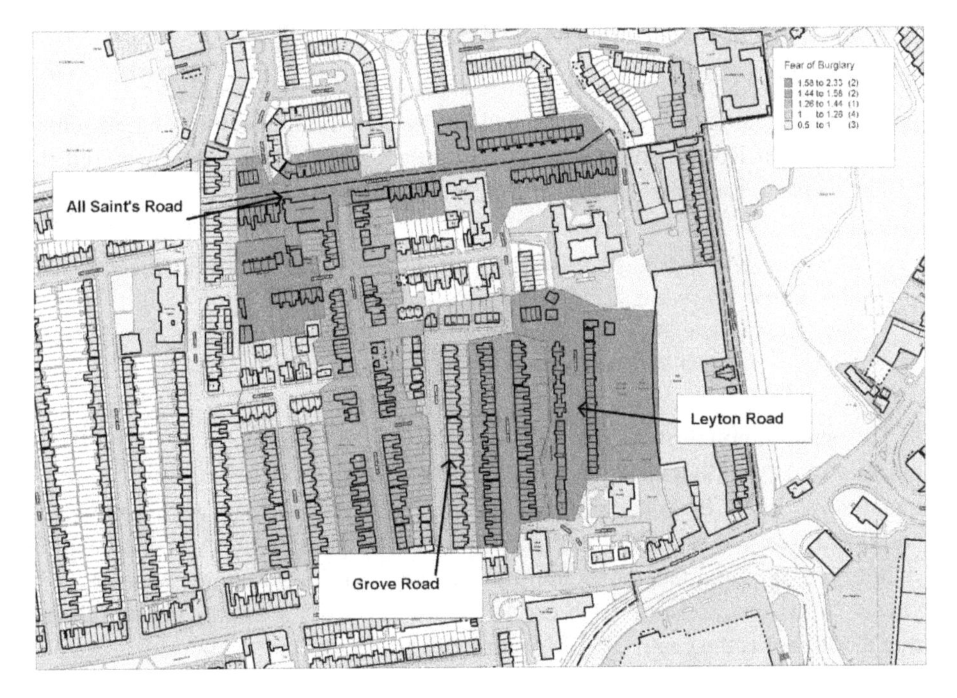

Figure 13.3 Fear of burglary in the Merton study area, London.

Haydon's Road is a main thoroughfare linking Merton with Earlsfield; it has a lot of traffic (both foot and vehicle) and this may contribute to an overall unsteady feeling. Wandle Bank is rather exposed, with the river forming one side of the road. Grove Road, however, is contained in the midst of the housing block, so why should it have a higher fear of crime?

Interpretations can come from the comments made by the residents in the free text section of the questionnaire:

> *'The street lighting is poor on Grove Road, it makes me feel unsafe at night'*
> *'The streets are dirty – rubbish bags are left out for days'.*

These two responses help to indicate, firstly, that poor lighting may contribute to fear of crime (as per Ramsey, 1991); and that environmental indicators can also contribute to fear (as per Innes and Fielding, 2002). This interpretation by the residents can now be used to drive action: by contacting the Street Management team of the Local Authority to examine street lighting and improve rubbish collection (perhaps in a problem-solving framework), we hope to have a positive effect on fear of crime in Grove Road.

13.4.2 Youth anti-social behaviour (ASB)

Fear of Youth ASB seems to be centred on the primary school (Figure 13.2), one possible explanation being that maybe older children use the facilities out of school hours. Hotham Road also appears to be a relative hotbed of fear. We asked the resident's panel for some clarification on this matter:

> *'Gangs of older kids hang around the primary school and play football on the field after school has finished. They graffiti the cars too.'*
> *'Kids mess around in the street on their way to and from the school'*
> *'Two kids in Hotham Road make our lives hell'.*

This small amount of insight helps greatly with the interpretation of the maps; essentially, key individuals act as 'interpreters' of the knowledge we have on a quantitative level. Both the primary school and Hotham Road were mentioned, and detail was added which the mapping could not possibly have afforded – thus reinforcing the importance of a multifaceted method. In these three short extracts we see 'kids' mentioned in each and the school in two – strengthening the theory that the school is the centre of the Youth ASB problem in the area.

13.4.3 Burglary

Burglary is traditionally high in these streets, and was indeed the initial motivation for carrying out the work. It is possible that the high fear in Leyton Road is due to an easy escape route and limited natural surveillance. All Saints Road again features highly (Figure 13.3); perhaps because it borders an area of different housing types – a 1970s' estate with lots of open spaces between the properties, which again could potentially make residents feel vulnerable. Houses on the outskirts of neighbourhoods are more likely to be burgled (Weisel, 2003).

A resident said:

> *'There were three burglaries on Grove Road in a week. A man wrote a message on his door to the burglars – so now we all know what's happening!'*

This is very interesting, as the Grove Road resident actively raised the fear of crime in his immediate vicinity (perhaps rightly so, given that there were three burglaries in a week); yet as we can see, Grove Road is in the middle ranking of the roads for fear of burglary. Perhaps there are some other factors more telling than the Grove Road story? Perhaps not everyone on Grove Road had as much access to the knowledge of the resident who fed us this information?

13.5 Methodological applications and considerations

With the advent of Neighbourhood Policing in London (and across the UK), an in-depth knowledge of levels of residents' fear and hotspots of those fears, for differing crime types, throughout a small urban geographical area can aid greatly in the reassurance process. This can be for tasking purposes (perhaps to send foot patrols down one road with a particular fear of crime), for publicity reasons (to send a positive reassurance message to a fearful area), or even for target-hardening schemes to help reassure an area with high fear of burglary.

Eventually a series of thematic maps could be made up for larger geographical areas (such as a police command unit area) where at a glance areas with high levels of fear could be identified. Further qualitative work could be done in those streets to further break this down; perhaps there is an overhanging bush or dark alleyway fostering fear of crime, which, with a simple application of multi-agency working, perhaps in a problem-solving framework, could be cut back or lit up, helping to further narrow the Reassurance Gap.

The method described in this case study is designed to provide a relatively simple framework for neighbourhood police officers or other local police/crime reduction teams. It demonstrates how, in Merton, easy-to-use graphical representations of the fear of crime at the street level can be produced. This is beginning to support a number of applications designed to better enable the targeting of resources to help reduce local residents' fears of crime.

13.6 References

Blears, H. (2005) Foreword. In *The National Community Safety Plan 2006–9*. London: Home Office.

Home Office. (2005) *The National Community Safety Plan 2006–9*. London: Home Office.

Ditton, J. and Innes, M. (2005) The role of perceptual intervention in the management of fear of crime'. In *Handbook of Crime Prevention and Community Safety*, Tilley, N. (Ed.). Cullompton: Willan Publishing, pp. 595–623.

Innes, M. and Fielding, N. (2002) From community to communicative policing: 'signal crimes' and the problem of public reassurance. *Sociological Research Online* **7**(2). http://socresonline.org.uk/7/2

John Howard Society of Alberta (1999) *Fear of Crime* (online). http://www.johnhoward.ab.ca/PUB/C49.htm

Ramsey, M. (1991) *The Effect of Better Street Lighting on Crime and Fear*. Crime Prevention Unit, Paper 29. London: Home Office.

TNS Social (2005) *Merton Annual Residents' Survey*. London.

Weisel, D. L. (2003) *Burglary of Single-family Houses*. Center for Problem-Oriented Policing. http://www.popcenter.org/Problems/problem-burglary-family.htm

Wood, M. (Ed.) (2005) *Merton Borough OCU Policing Plan 2005/6*. London.

14 NightVision – visual auditing of night-time economy related incidents in Bath and North-East Somerset

Jon Poole

14.1 Introduction

The NightVision project was an initiative to help to better understand the geography and nature of the night-time economy (NTE) in Bath city centre. Alcohol related crime and disorder had been identified as a development priority in the 2004 Community Safety and Drugs Partnership 'Community Safety and Drugs Audit'. The project formed part of a long-term strategy to provide greater understanding of this problem.

This case study presents a brief methodological overview of the project and assesses the results, and in so doing highlights some of the key issues involved with conducting primary research in a Crime and Disorder Reduction Partnership (CDRP). To date there have been three phases of the project; two in the winter–spring period of 2005 and 2006, and one conducted during the summer months of 2006 to provide an analysis of seasonal variations.

14.2 Project design and implementation

From the outset of the project, a multi-agency steering group was formed; this group included a wide range of internal and external partners, including representatives

Crime Mapping Case Studies: Practice and Research Edited by Spencer Chainey and Lisa Tompson

of local licensees and residents groups. Collectively, it was their role to oversee, design and implement the project. This was an effective partnership from the outset, and one which had the full support of the Police's Community Safety Team. They proved invaluable in ensuring that the project was conducted safely and efficiently. The group was primarily responsible for:

- drawing up the routes that would be walked, and the audit categories under which data would be recorded;
- risk assessing the routes that would be walked;
- recruiting volunteer surveyors for all four nights of the project.

14.3 Methodological considerations

14.3.1 *Recording incidents*

The primary basis for the methodology was the Living Streets' Community Street Audit (www.livingstreets.org.uk/what_living_streets_do/cs_community_street_audits.php) which attempts to provide a dynamic methodology for understanding and classifying individuals' experiences in the physical environment. One innovative aspect of the project was that it sought to add to the qualitative nature of the *Living Streets* model, in creating a geographically referenced body of quantitative data from a visual audit.

The Community Street Audit advice suggested that there should not be a large number of categories recorded. In defining these categories the principles of signal crimes were used. Based on research conducted by the University of Surrey (Innes and Fielding, 2002) and adapted by the Association of Chief Police Officers, signal crime principles suggest that viewing and experiencing crime and disorder has a long-term disproportionate impact on individuals' perceptions of safety and fear. Comments made by surveyors could be seen to corroborate this theory.

> *'Generally places with . . . character/interest . . . are where I feel safe, regardless of time of night . . . But any swearing or shouting . . . quickly makes me feel uncomfortable'* (Bath University Student surveyor, NightVision 2, night four).

It was understood that by walking through the environment, surveyors would not experience a static version of the city centre environment. As a result it was necessary to break incidents down into two categories. The first related to behaviour, i.e. actions that were observed by surveyors as they occurred, and the second related to physical features, i.e. categories and features *in situ* in the street. The categories were therefore chosen on the grounds that they would represent situations where alcohol related

Table 14.1 Categories of behaviour and physical features to be recorded by the surveyors

Behaviour categories	Physical feature categories
Aggression or intimidating behaviour	Dropped litter/food
Nuisance noise	Bottles, cans or broken glass
Obviously intoxicated persons	Urine and vomit
Urinating and vomiting	Criminal damage (e.g. to cars, street furniture, including vandalism)
Street drinking	
Violent criminal behaviour	

disorder could arise, had arisen or was in the process of occurring. These categories are listed in Table 14.1.

Owing to cost restrictions and the practicalities of recording issues, a laminated sheet and marker pen were used by the surveyors, who recorded what they saw using an index code (see Figure 14.1 [4.8]).

Analysis of the data that examined the location of incidents recorded on the same night and the same route showed that some categories experienced notable fluctuation

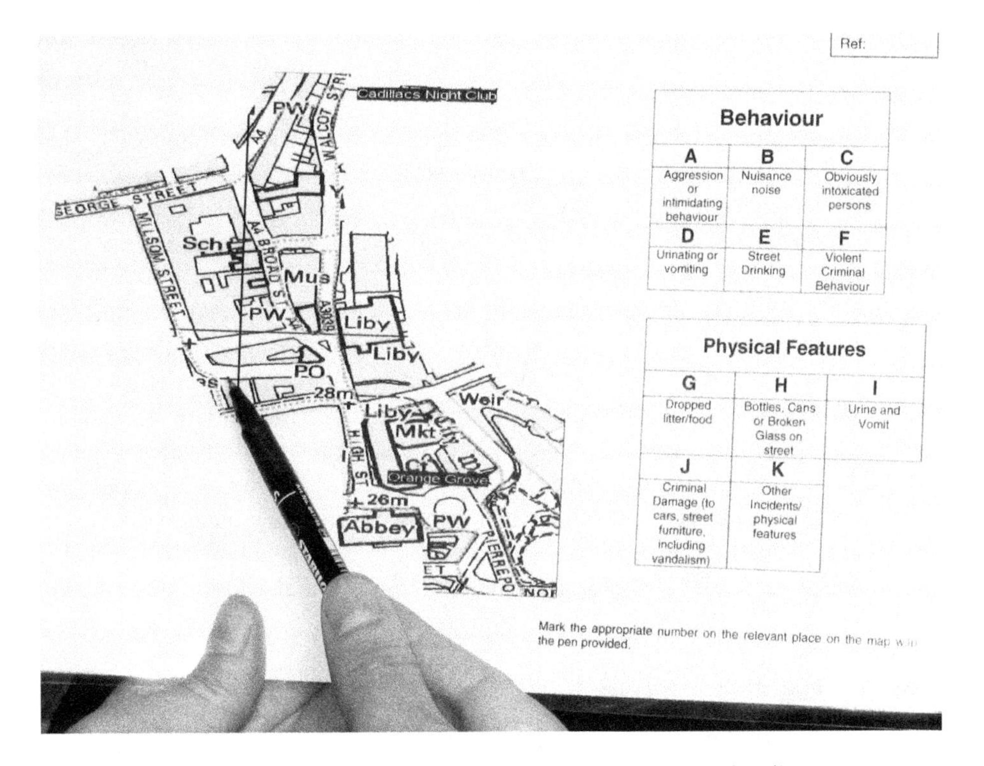

Figure 14.1 The survey form used in the NightVision visual audits.

between locations. The most notable in terms of their variation were 'aggression and intimidating behaviour', 'nuisance noise' and 'obviously intoxicated persons'. Despite this variation, however, across the studies a degree of consistency could be observed in terms of the geographical location of these incidents. The remaining categories were roughly consistent between surveyors.

14.3.2 The surveyors

Surveyors were chosen as representatives of specific 'interest' groups in the NTE, notably residents (those who experience it without necessarily engaging with it), users (those who are out in the NTE), professionals (those who have a financial benefit from the NTE) and finally a group which could be seen as a more objective 'other'. In addition, it was seen as important to gain a demographic balance compared with the population as a whole, particularly in terms of age, gender and ethnicity. In practice, however, the under 30 age range was most difficult to engage.

Because of the practical difficulties of recruitment, the following groups were targeted; students (representing users and a large proportion of the under 30 demographic), residents, Licensees and Health Care Professionals (seen as the more objective 'other'). The groups were of equal number and one representative of each group conducted the research on each night.

There was no evidence to suggest that the 'themed surveyor' approach taken affected the research outcomes. One significant gain from this approach, however, was the inclusivity of all sides of the argument (i.e. alcohol-related disorder) in gathering the data, which assisted in increasing the appearance of objectivity when presenting the results in a multi-agency forum.

Due to health and safety considerations, it was decided that surveyors would have to be accompanied by some form of security. Security professionals were appointed to the project, and they were responsible for delivering a safety briefing for all surveyors and discretely accompanying surveyors on all their route walkabouts. Staff other than police officers were chosen to shadow the paired surveyors because of the need to keep as much of a physical presence with volunteers as possible. Operational police officers would have been required, had any incidents occurred, to instantly attend those incidents, thus potentially leaving surveyors stranded in dangerous positions, or proceeding on their auditing without support.

14.3.3 Geography and logistics

The geography of the routes was based entirely on existing hotspots of crime. An analysis of multi-agency data was undertaken to provide the location of key hotspots around which the analysis could be conducted. Datasets used in this analysis were:

- police recorded crime data;
- public complaints to the police regarding nuisance and disorder;
- local CCTV data;
- local ambulance data.

Accident and Emergency data were also gathered, although these data did not contain reliable incident location references and so were not included.

The route, when plotted, was broken down into two routes, one route was deemed unsuitable as it could not fulfil the following considerations:

- the route should take no longer than 20 minutes to be walked at an average pace (the routes, when walked by surveyors, took between 30 minutes and 45 minutes to cover, given that it was not always possible for surveyors to note comments or observations while walking);
- the route needed to be circuitous;
- the route needed to follow fairly obvious street patterns (to avoid surveyors getting lost or concentrating on the map rather than the environment).

Figure 14.2 demonstrates how the routes were aligned to known hotspot areas.

14.3.4 Timetabling and dates

The background historical information gathered at the outset allowed definition of 'average' days on which to conduct the project. Similarly, it allowed for the location of temporal hotspots in order that accurate recording of incidents could occur.

The routes were walked at the following times: 23:15, 00:15, 01:15 and 02:15 hours. There was a margin of error of approximately 15 minutes for each route, dependent on the speed surveyors walked.

14.3.5 Analysis

All incidents recorded on the forms were transposed into digital format to allow for analysis. Each incident was ascribed a unique reference including route, night and surveyor reference. This allowed for the removal of incidents duplicated by surveyors recording the same incident.

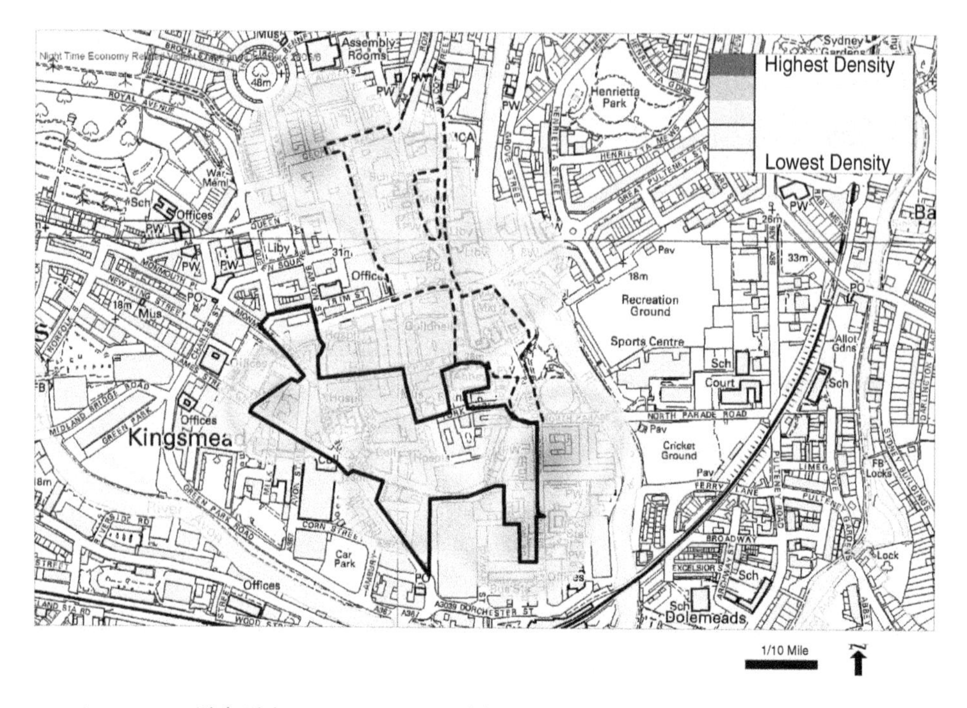

Figure 14.2 NightVision survey routes and how they were aligned to existing hotspots.

Geographical analysis was conducted on all categories where a suitable number of incidents had been recorded to identify if concentrations of observations were present. In addition, overviews of rates and trends by date, time and surveyor category were conducted. Analysis was conducted primarily using Microsoft Excel, MapInfo Professional and MapInfo Vertical Mapper. An example of this output is provided in Figure 14.3. In light of the audience for the documentation, simplistic presentation of the data was seen to provide greater clarity.

14.4 Findings from the NightVision surveys

Increases in incidents were experienced between the studies conducted in the winter–spring period and the study conducted in the summer months. These increases were expected and in line with anecdotal and scanned intelligence gathered from practitioners and crime and disorder trend data. 'Intoxicated persons' and 'nuisance noise' were consistently the most commonly recorded categories across all the studies. The surveyor groups had limited impact on the results and the most significant variations in the results were seen to relate to the individuals conducting the survey, rather than the specific night, route or time.

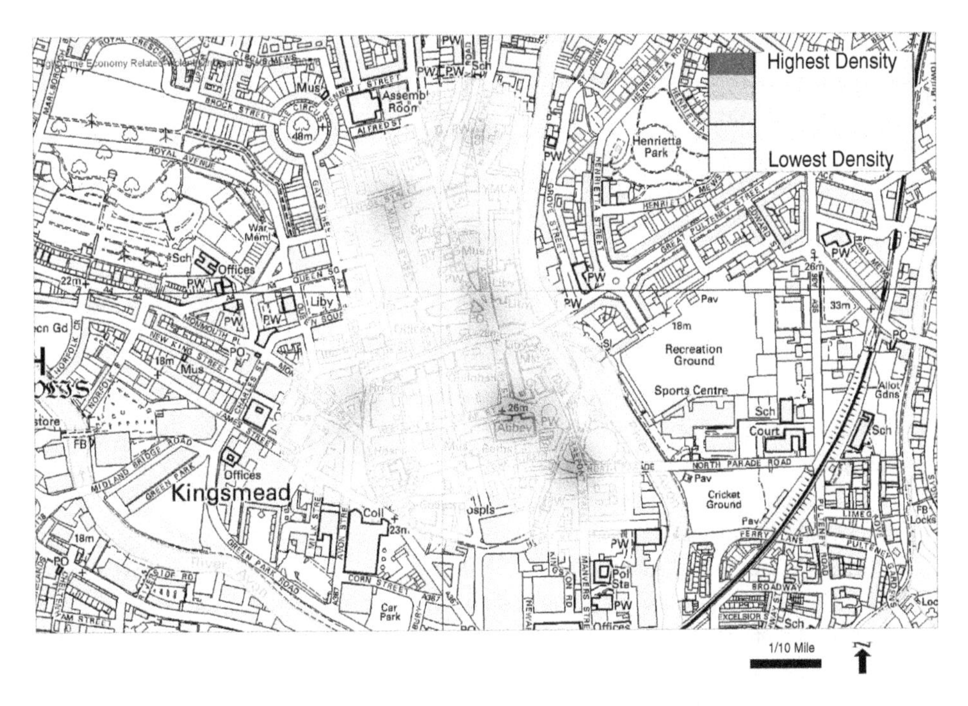

Figure 14.3 A sample of NightVision output analysis – hotspot analysis for obviously intoxicated persons (NightVision 3).

There were no statistically significant spatial correlations between the location of recorded incidents of crime and disorder and the location of any recorded categories. However, visual analysis demonstrated that there was some consistency in location between specific hotspots in the city centre across a number of categories. These locations were consistent with the placement of late night fast food venues and a number of larger late night licensed premises. In the vast majority of cases surveyors felt safe or very safe while out in the city centre at night, although the presence of the security professionals was believed to have some impact on this perception.

Through qualitative analysis of the de-brief documents it was possible to identify commonalities in issues which were seen to have a notable impact on surveyors' perceptions. Across all the studies, the most notable issues were those relating to visible uniformed presences, either in the form of police officers, door staff or other security personnel. These features, however, only had a positive impact on perceptions when they were well managed. Comments were received regarding door staff behaving in a manner that reduced feelings of safety in the surveyors which reinforced this supposition. Environmental issues, such as lighting and refuse collection, were seen to have had notable impacts on surveyors' perceptions, with waste accumulations in particular being noted as reducing feelings of safety in an area. This is exemplified in the following quote:

> *'The steps leading down to ##### Street felt unsafe ... you felt like you were walking through a slum'* (Licensee surveyor, NightVision 3, night four).

14.5 Generating action

Driving action based on the evidence of the project had been a fundamental objective from the outset. There was strong support for the work within the local CDRP, and with the assistance of senior officers from a variety of agencies the report was presented to a wide range of groups and agencies. These included local Licensee groups; Police and Partnership tasking groups and Local Authority Overview and Scrutiny panels. Tactical decisions were usually made within the first three months of each study being produced, with the location analysis providing context for operational police tasking and also for identifying licensed premises which may be contributing to recorded incidents.

Time scale was one of the key hurdles to delivering strategic action from the project, with over thirteen months elapsing before any more strategic outcomes of the research could be defined. The most fundamental outcome was the use of the research as the primary evidence for the creation of a night-time cleansing team in Bath city centre. The evidence has also been submitted as part of a multi-agency application for a Cumulative Impact policy for the city centre area.

The report generated a large amount of local media attention. Careful management of the messages contained within was required to ensure that the publication of the report itself did not contribute to reducing the general public's perception of the city centre at night.

Although issues of internal reliability have meant that it is not possible to accurately use the tool as a method of tracking incidents longitudinally, the work is now being taken in a direction towards regeneration-based outcomes, particularly in terms of street lighting and urban design.

14.6 Conclusions

The NightVision projects have been successful in being accepted by diverse stakeholders as a methodologically reliable method of assessing a wide range of issues in the night-time economy, subsequently generating action. The appearance of methodological neutrality has proved as important as the results themselves in achieving this aim.

In addition, the variations in individual perception have proved to be one of the key findings of the report, as the following comments, received from two individuals who walked the same route at the same time, demonstrate:

> *'(I felt) fairly unsafe, late Bath is not the place it used to be ...'* (Bath Resident surveyor, NightVision 1, night three).
> *'Genuinely the most quiet and unintimidating evening I have ever spent in Bath'* (Student surveyor, NightVision 1, night three).

As a result, the study has particularly shown that accepting these variations and the associated compromises required is key to the successful management of a night-time economy.

Finally, although the need in CDRPs for well funded and organised primary research is well documented, for such research to drive strategy and decision-making there is an enormous dependency on a willingness to change from partner agencies and groups.

14.7 Reference

Innes, M. and Fielding, N. (2002) From community to communicative policing: 'signal crimes' and the problem of public reassurance. *Sociological Research Online* **7**(2). http://socresonline.org.uk/7/2

Part V New techniques

15 The near-repeat burglary phenomenon

Derek Johnson

15.1 Introduction

Academic research in recent years has explored the spatial and temporal elements of residential burglary in the UK as well as in other countries (Bowers *et al.*, 2004; Johnson and Bowers, 2004). This research has concluded that burglaries cluster in both time and space. Where this is apparent, the risk of burglary is 'communicable' in a similar way to disease, that is it quickly spreads to other residential properties nearby but as time passes the risk of communication reduces.

This 'near-repeat' phenomena, as it has been dubbed, would appear to potentially contribute to supporting a proactive impact on crime reduction initiatives. It was on this basis that analysts in the Bournemouth Division of Dorset Police (on the south coast of England) saw opportunities for such strategies around residential burglary crime in 2005. This was at a time when the division was faced with tough burglary reduction targets and when it appeared to offer the ability of introducing a predictive capability towards directing reduction interventions.

15.2 Near-repeats in Bournemouth

Such potential predictiveness does, however, require a quick response. Research indicated that properties close to an initial residential burglary event were at their highest risk of attack within 48 hours of that initial event. Although risk remains high for a further five days before tailing off over 28 days, the challenge was to introduce a workable intervention that could respond almost immediately following a trigger event. If processes could be found to utilise this identification of high crime risk areas, the process of identifying near-repeats could empower us to map risk with

Crime Mapping Case Studies: Practice and Research Edited by Spencer Chainey and Lisa Tompson

good integrity and begin to predict where offences were most likely to occur in the future. Both a time (two days, seven days and 28 days) and distance factor (400 m from initial event) had been determined, allowing us to focus towards specific areas that could receive the intervention after the initial event.

The second, and perhaps most important challenge, was to establish if the residents of Bournemouth were actually experiencing the phenomena. If this was the case then where in Bournemouth was it happening? Another factor learnt from the research was that near-repeats are not all encompassing, but appear to be restricted to certain areas of a conurbation.

The Jill Dando Institute of Crime Science undertook analysis of 12 months of burglary data for Bournemouth and established that, on a whole Basic Command Unit (BCU) scale, the near-repeat phenomenon was indeed present. The analysis also confirmed that the high risk features were over distances of 200 m and at time scales of two (highest risk), seven and then 28 days. Bearing in mind that the data analysed were for the entire BCU, the results were very promising. There was clear evidence that the near-repeat phenomena had been present in the town during 2004 but that, unlike the situation found elsewhere, the near-repeat distance was only 200 m (and not 400 m as shown in previous studies – Bowers *et al.*, 2004).

No quick and easy method was available to identify the particular housing areas suffering from near-repeats, although further research findings could be taken into account to reduce the problem of having to analyse data over the entire residential geography of Bournemouth. For instance it was known that there was a very definite negative correlation between repeat residential burglary offences and near-repeat offences; that is near-repeat offences did not tend to occur in areas that suffered from high numbers of repeat offences. Given this information it was perhaps unsurprising to also learn that near-repeats did not tend to be manifest in areas of high social deprivation either.

15.3 A methodology for analysis and action

To establish communities in which near-repeat offences were manifest local analysts had to extract residential burglary data for 12 months. This was then plotted within a geographical information system (GIS) and separated into new tables depicting offences that had occurred in fortnightly blocks over the year long period. Having plotted the first two weeks of crime, 200 m buffers were created around each crime location and the next two week block was plotted. This process was then repeated for each fortnightly block of data. By simply visualising the data in this way, but never having more than two fortnightly blocks of data visible at any time, it was possible to identify those burglary crimes from the second fortnightly period that lay

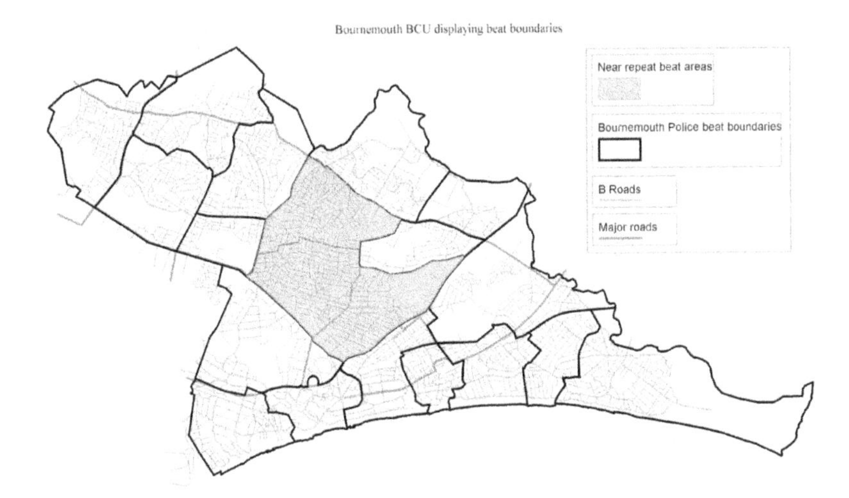

Figure 15.1 Police beat areas in a Basic Command Unit (BCU) of Bournemouth where the near-repeat burglary phenomenon appeared to exist.

within the 200 m buffers around the incidents from the first fortnight. By repeating this process for the whole year it was possible to establish three police beat areas where the phenomena had been operating almost throughout the beats. All three beat areas were contiguous and two, unusual for police beat areas, actually had boundaries that appeared to be useful for the purpose of near-repeat analysis, in that they followed significant roads in the area and therefore tended to delineate between different communities. These beat areas are shown as shaded in Figure 15.1.

It was decided to develop a strategy that would operate in just two of the three beat areas for a pilot period. There was some evidence of near-repeats in other areas of the town but these were within beat areas rather than encompassing them as a whole. Although police beat areas tend often to be drawn arbitrarily (and those in Bournemouth do not reflect other geographies such as local authority ward areas) they are defined by police officers, for police officers and are therefore understood by police practitioners. This was an important factor as it was clearly going to be these practitioners that would be delivering any reduction strategy.

Delivering 'hard hitting' crime prevention advice to properties within 200 m of a burglary event was seen as paramount in a bid to reduce burglary within the selected beat areas. An intervention strategy was therefore put into place that hinged on three key factors.

1. The ability of analysts to identify the locations of dwelling burglaries within the two chosen beat areas, and wherever possible within 24 hours of the initial event.

2. The ability of Police Community Support Officers (PCSOs) to respond to the analysts' recommendations on the same day as a high risk location was identified.

3. The delivery on a face to face basis of some hard hitting crime reduction advice to those addresses identified by analysis as being at high risk. This was to include householders being told about the near-repeat phenomena and it being made clear that police had reason to believe that their properties were at increased risk of attack.

15.4 Delivering a near-repeat intervention

The police crime reduction intervention began on 1 April 2005 and was preceded by a number of press releases. This was an intervention that the local media found particularly appealing and as a result were keen to follow over time. This was found to be advantageous as the main audience for the intervention were to be those that use and respond to the local media.

Analysts responded by identifying all burglaries in the selected police beat areas within 24 hours of reporting (wherever possible) and preparing detailed maps indicating the dwellings that were at the highest risk of imminently suffering a future burglary. This utilised a simple process of drawing a 200 m buffer around the location of the trigger event on a conurbation-level map. By then using further findings from the original research, such as houses on the same side of the road as the trigger event being at higher risk than those on the opposite side, actual buildings were highlighted on the analyst's map as first, second and third priority requirements.

These maps were passed to PCSOs who were priority tasked with visiting as many of the high risk addresses as possible within that day, and then similarly calling at the second and third priority addresses until all had been visited. An emphasis was placed on visiting the high priority addresses as soon as possible. At all times 'one on one' communication with householders was sought, but ultimately all identified addresses received a crime reduction information pack together with correspondence describing the initiative and the increased risk factor. By January 2006 over 3000 dwellings had been visited by PCSOs at various times, all receiving specific dwelling burglary crime reduction advice.

15.5 The impact

The two beat areas that received this focused attention experienced a dwelling burglary reduction of 37% and analysis was undertaken to further evaluate the success or otherwise of the action taken.

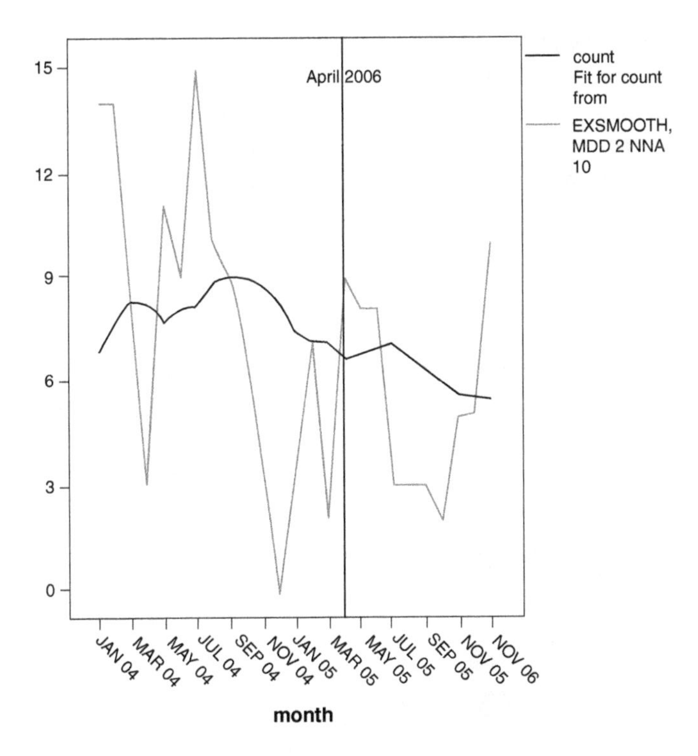

Figure 15.2 Changes in burglary dwelling offences in Bournemouth between January 2004 and January 2006

With a start date of the 1 April 2005, two 10 month periods of data ('before' and 'after') showed that dwelling burglary crime had reduced by 32.5% in the 2005–06 period. A time-series chart (Figure 15.2) indicated that burglary dwelling offences in the area had been reducing from about October–November 2004 in any case. This can be seen in the exponentially smoothed line in Figure 15.2. It was therefore necessary to establish if burglary crime had fallen at a greater rate than could have been expected from the ongoing downward trend.

In order to identify a further area of the division which displayed similar demographic qualities to the intervention areas, and could therefore be used as a control site, 2001 National Census data relating to premises of a similar type were sourced. This process identified twenty Census Output Areas displaying similar rates of ownership and had similar levels of burglary dwelling offences during the intervention period. Twenty census output areas were also found in an area to the south that displayed similar rates of occupier mortgage ownership and where the near-repeat phenomena was known to be apparent. This area had not been subject to intervention and was separated from the intervention area by about four miles.

Comparisons were then made between the two burglary crime rates. Figure 15.3a displays a bar chart of the mean values of dwelling burglary crime rates in the two areas both before and after the intervention began. From this we can see that burglary crime did fall in both areas and possibly more so in the 'action' area. The possibility of a seasonal factor was considered but did not appear to be apparent. Figure 15.3b shows certain descriptive statistics for both areas before (controlpre and actionpre) and after (controlpost and actionpost) the targeted interventions.

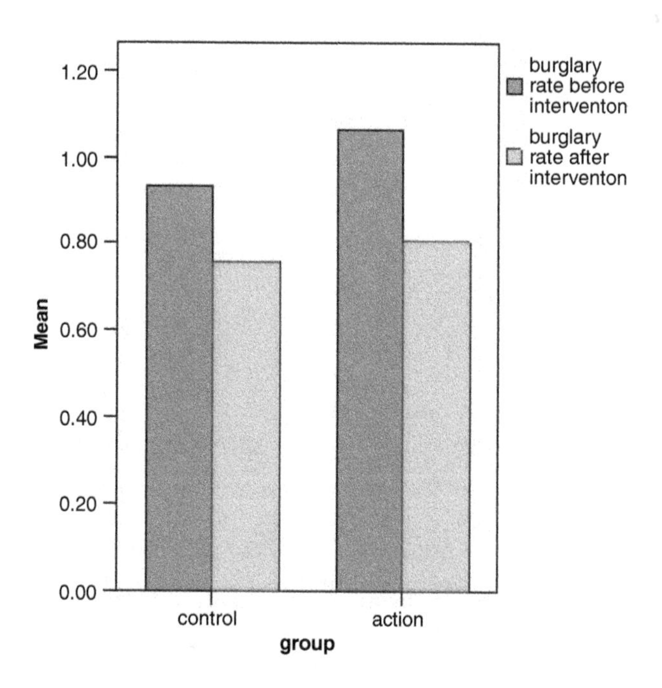

Descriptive Statistics

	Range	Minimum	Maximum	Mean	Std. Deviation
controlpre	1.17	.58	1.75	.9320	.39869
controlpost	**1.56**	.19	1.75	**.7580**	**.50736**
actionpre	1.57	.20	1.77	1.0620	.50920
actionpost	**1.18**	.39	**1.57**	**.8060**	**.41979**

Figure 15.3 Changes in burglary dwelling rates in the action area and the control area: (a) displayed as a bar chart and (b) as descriptive statistics.

The following points (shown as bold in Figure 15.3b) are of note from the descriptive statistics:

- The post-intervention range (maximum–minimum) in the control area was higher than the pre-intervention range, unlike the action area where the range reduced.
- The maximum value in the action area fell following the intervention, whereas in the control area the value did not alter.
- The mean burglary rate in both areas fell following the intervention although the fall in the action area was considerably greater than in the control area.
- The standard deviation in the control area increased after intervention.
- Although the action area experienced a reduction in burglary dwelling that equated to about 5% greater than the control area it could not be shown as statistically significant.

These descriptive statistics offer corroboration that burglary dwelling crime fell in both areas and that there were marked differences between the 'before' and 'after' values. This indicates that the decrease in dwelling burglary crime was different in both areas and that possibly the dynamics of those reductions were also different.

Dwelling burglary data for the action area over the same 20 month period was also examined on a spatio-temporal basis and it was the results of that analysis that were perhaps the most telling feature of the intervention strategy. It was concluded that since the intervention was put into place the 'near-repeat' phenomena was no longer apparent.

By mapping and treating residential burglaries in areas known to encourage a spatio-temporal link between such crimes as indicators of forthcoming risk, action had been taken that had led to a change in the spatial behaviour of offenders. This spatial change can also be visualised with the use of standard deviational ellipses of burglary crime (produced using the program CrimeStatIII (Levine, 2004)). Figure 15.4 displays a background map with the two shaded areas representing the beat areas subject to the intervention. Standard deviational ellipses relate to the residential burglary crime in those intervention areas before and after April 2005. Noticeable is the quite acute change in direction of the ellipse.

15.6 Conclusions

The use of the near-repeat phenomena in the Bournemouth BCU has been ongoing. In terms of residential burglary intervention, the strategy has been expanded into further beat areas where near-repeats were found to occur. A postal questionnaire to residents

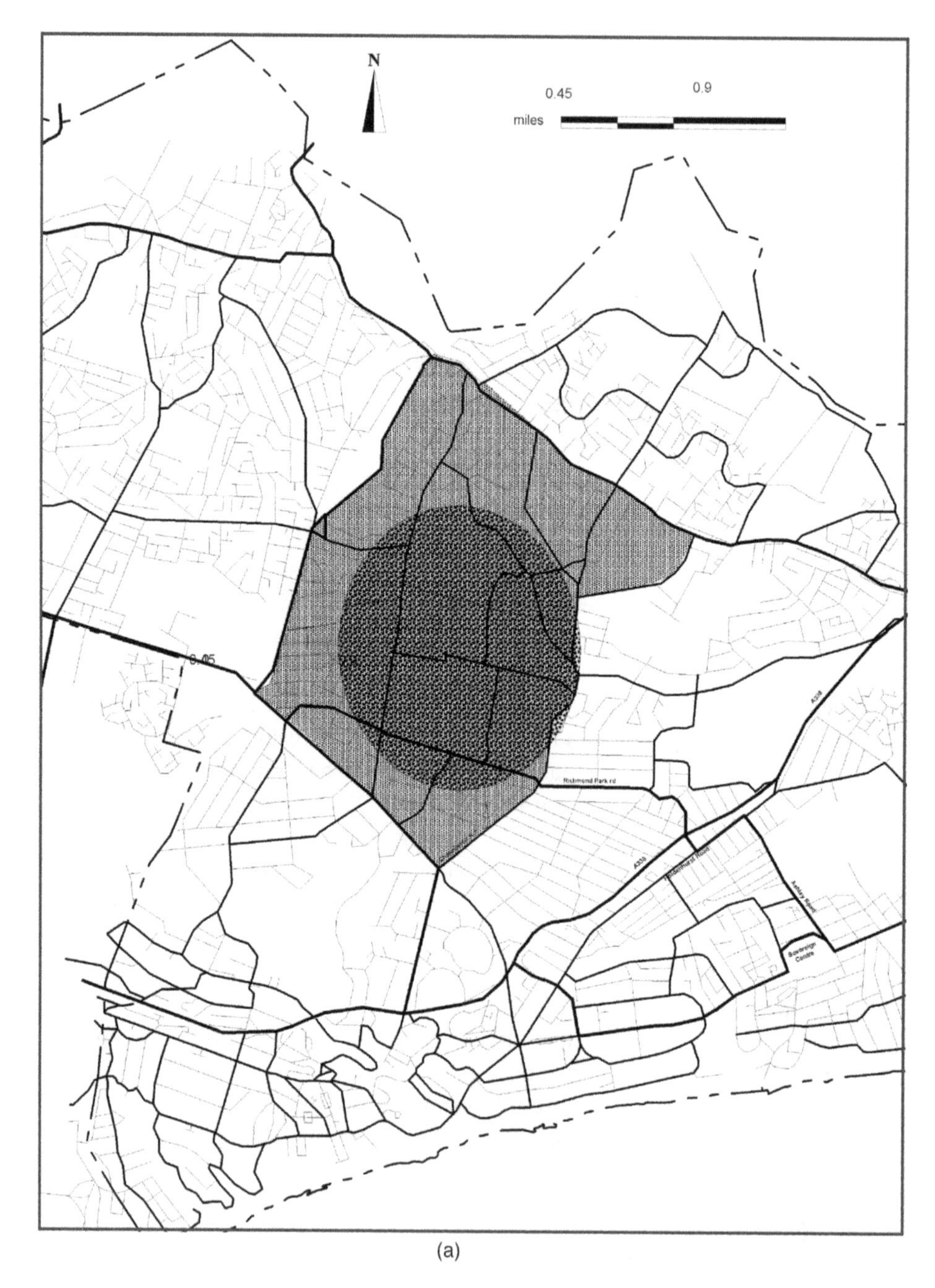

(a)

Figure 15.4 Standard deviation ellipses of burglary dwelling crime in Bournemouth (a) before the crime reduction initiatives and (b) after the crime reduction initiatives.

(b)

Figure 15.4 (*Continued*)

also identified overwhelming support for information to be passed concerning local crime, with very little evidence of any resulting increase in the fear of burglary crime.

Speed is, however, of the essence as the high risk window is small in temporal terms. It is therefore essential to develop both analytical and operational processes that can react both quickly and efficiently. Interestingly there was one short period during 2006 when the BCU was unable to provide the operational resources to react to the prospective mapping. Although anecdotal only, it was noted that in that period alone a number of near-repeat burglaries were recorded in the intervention areas. In order to reduce the time between identification of a trigger offence and communicating the risk to residents', mobile telephone text messaging technology is now being utilised. By using a web-based service and asking residents to register with their mobile phone numbers and postcodes a generic text message is sent to registered residents in high risk postcodes when a trigger burglary event is identified. Although these addresses are still visited, this does allow for speedier communication with residents, particularly those who may be working during the period when PCSOs are visiting.

Steps are now underway to identify areas in the BCU where the same spatio-temporal clustering can be found in thefts from motor vehicle crime. At this stage a number of beat areas have been identified that display evidence of such clustering. It is intended to use recorded theft from motor vehicle offences in such areas as trigger crimes in the same manner as described for burglary dwelling. However, unlike the burglary initiative, the existence of a trigger crime will be used to create enforcement opportunities, primarily by prioritising the deployment of 'trap' vehicles within the identified high risk zones. Again, the highest risk window is small at no more than three days and the distances from the initial event remain at about 200 m. It is hoped that such a strategy will lead to an enhanced predictive mapping capability that can influence the use of enforcement resources to best effect and shape tactical responses. Dorset Police is committed to continuing to develop an approach to mapping crime risk by utilising research in this area, and so developing prospective crime mapping to intelligently lead crime reduction and enforcement opportunities.

15.7 References

Bowers, K. J., Johnson, S. D. and Pease, K. (2004) Prospective hot-spotting: the future of crime mapping? *The British Journal of Criminology* **44**(5): 641–658.

Johnson, S. D. and Bowers, K. J. (2004) The stability of space–time clusters of burglary. *The British Journal of Criminology* **44**(1): 55–65.

Levine, N. (2004) CrimeStat III: a spatial statistics program for the analysis of crime incident locations. Houston, TX: Ned Levine and Associates. Washington, DC: National Institute of Justice. Available at http://www.icpsr.umich.edu/nacjd/crimestat.html

16 Simulating crime to inform theory and practice

Elizabeth Groff

16.1 Introduction

Gaining a better understanding of crime and how to prevent it is a challenging endeavour. To better understand how a crime occurs requires a variety of information that is not routinely available. For example, to differentiate situations in which a crime occurs from those in which it does not, would require us to collect information about both types of situations. Alternatively, to understand why crimes occur where and when they do, a variety of individual-level data is needed that describes the places, people and situations involved in crimes. The inability to collect the individual-level data necessary to characterise human travel behaviour in general and the situational elements of crime events in particular, however, is an ongoing barrier (Huisman and Forer, 1998; O'Sullivan and Haklay, 2000). Primarily due to cost and privacy concerns, information about the routine movements of citizens across space and time is not collected. Even if the data could be obtained, another hurdle to answering questions about complex phenomena such as crime is finding modelling tools capable of capturing the dynamic nature of human activities and interactions of individuals when they converge.

In response to these challenges, some researchers are investigating the use of simulation models (Brantingham and Brantingham, 2004; Brantingham and Groff, 2004; Eck and Liu, 2004). Simulation models involve the creation of an artificial society that exists in a computer program. This representation is usually simplified and captures only the most important elements and relationships of the phenomenon being studied. One well-known example of a simulation model is the computer game SimCity® (Electronic Arts, Inc., Los Angeles, CA) . In this computer game, players

Crime Mapping Case Studies: Practice and Research Edited by Spencer Chainey and Lisa Tompson

make decisions about developing the city and then are able to 'see' the consequences of those decisions as they unfold. Simulation modelling is a more formal version of this process that allows for the development of models based on theory and for the systematic testing of those models.

16.2 Agent-based modelling

Agent-based models are a type of simulation model in which important elements are represented by agents. An agent can represent a person, an organisation or even a neighbourhood. Each agent has characteristics and behaviours. For example, a person agent might have characteristics such as income, race, gender and age. The person might also have behaviours such as the ability to travel between locations or to make the decision to commit a crime. Each agent can act autonomously, making decisions based on the situation. Thus an artificial society populated with agents has the capability of simulating the dynamic elements that characterise human behaviour as well as the interactions of humans with each other and with their environment.

Several characteristics of agent-based models directly address the challenges facing earlier crime research and suggest agent-based modelling (ABM) as an important component of a new, more flexible methodology. Reasons for this include:

- ABM allows heterogeneity among individuals that more closely approximates the variety found in life.
- ABM is better able to accommodate the non-linearity in relationships that is frequently evident in complex and dynamic interactions (Epstein and Axtell, 1996; Gilbert and Terna, 1999; Dibble, 2003).
- Individuals in agent-based models are able to make dynamic decisions based on changing information (Bonabeau, 2002).
- ABM enables the execution of systematic experiments using simulation (i.e. being able to hold the agents and/or the landscape constant and then vary one or both of them systematically). This feature provides a level of control difficult to attain using traditional social science methods.
- ABM provides a cost-effective and ethical methodology for evaluating theories prior to empirical research. Simulation experiments are cheap compared with empirical research such as randomised experiments. In addition, they are not subject to the ethical concerns often encountered when human subjects are involved.

The use of ABM in the social sciences has increased over the past ten years (Gilbert and Troitzsch, 1999; Macy and Willer, 2002). Noting the critical role that the interaction between human behaviour and the environment plays in social phenomena, some researchers have called for the use of realistic rather than artificial landscapes (Albrecht, 2005; An *et al.*, 2005; Brown *et al.*, 2005). Although traditional simulation is well-suited to keeping track of time and dynamic interactions, it is not so good at incorporating spatial and environmental elements. Thus to better model dynamic events such as crimes, a combination of both ABM and geographical information systems (GIS) is needed. Recent advances in technology have enabled the creation of the software packages necessary to permit agents to be 'situated' in a particular spatio-temporal milieu (e.g. agents can travel along real streets and respond to the characteristics of a real environment). Agent Analyst is a concrete example of software that links a GIS, ArcGIS (ESRI, 2005), and an ABM package, RepastPy (North *et al.*, 2006). The combination of the strengths of ABM and GIS is necessary in order to move away from the use of artificial landscapes and instead model individuals in their environment.

These technological advances provide the tools for addressing the limitations of previous research. They facilitate the creation of a GIS–ABM model that is capable of capturing and analysing:

- the process involved in the convergence of offenders, victims and guardians at a particular place and time;
- the interactions that take place;
- the culmination of those interactions in the form of emerging crime patterns.

Models of crime can take on a variety of purposes. At the outset, models are usually either created with the goal of accurately predicting patterns of crime events or to experiment on theory. Models created to predict crime are then tested against empirical crime patterns through a process called model validation. When models are used to experiment on theory, they are developed using a particular theory's view on the way the world works. They are evaluated by testing whether the model outcomes match what the theory would predict.

The model of street robbery developed here is based on routine activity theory and provides a better understanding of how the spatio-temporal aspects of human activity influence the incidence and distribution of street robbery events in Seattle, Washington. Accordingly, the point of this research is not to predict the pattern of street robbery events, but rather to operationalise the assumptions of routine activity theory in an artificial society and then test whether the model outcomes match the predicted outcomes.

The approach taken here emphasises theory testing but still in a theoretical world. In this way, the method represents an interim testing ground between the verbal formulation of the theory and the testing of theory with empirical data. Although this exercise does not result in a determination of whether a theory is true in the real world, it does provide a way to test the plausibility of the theory's assumptions. To this end, the most efficient (i.e. simple but effective) model possible is created and run, and the results subjected to rigorous testing.

16.3 Creating a theoretically based simulation model to test routine activity theory

This section provides an example of how a simulation model (i.e. an artificial society) can be developed based on a criminological theory, in this case routine activity theory (Cohen and Felson, 1979). Routine activity theory has been widely applied in empirical research and the theoretical framework is well-developed. Its focus on the situation in which a crime occurs shares characteristics with situational crime prevention and other opportunity theories of crime. The routine activity approach to studying crime holds that criminal acts require the convergence in space and time of likely offenders, suitable targets and the absence of capable guardians against crime. Changes in social structure have an impact on the frequency with which these elements converge by modifying the routine activities of offenders, victims and potential guardians. It has been hypothesised that if the frequency with which these elements converge in space and time increases, crime will also increase, even if the supply of offenders or targets remains constant within a city. The straightforward and concrete nature of routine activity theory makes it immediately applicable by law enforcement practitioners and amenable to exploration using a spatially informed agent-based model.

The methodology consists of the following three components (more details on the conceptual and implementation models are available in Groff (2007)). The experimental results are reported in full in Groff (2008, in press).

1. The theoretical constructs of routine activity theory are translated into a conceptual model (Figure 16.1). A conceptual model is a blueprint for the model that shows the important elements and the relationships between the elements. From this conceptual model three versions of a basic GIS–ABM model of street robbery are developed so that the dynamics of individual level decision-making and behaviour that produce macro-level street robbery patterns can be represented.

2. The role of geography (i.e. activity spaces) is explored. The agents interact in the 'real' environment of Seattle, Washington.

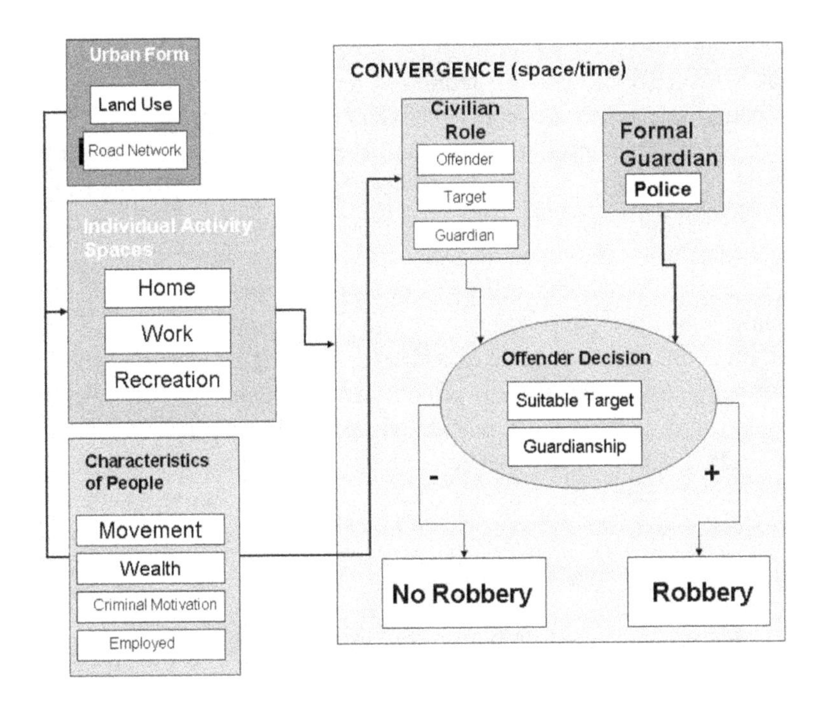

Figure 16.1 Conceptual model of street robbery according to routine activity theory.

3. A series of experiments are undertaken to test whether: (a) the theoretically predicted outcomes from routine activity theory match the model-produced outcomes; and (b) changing the spatio-temporal schedules of individuals produces different quantities and spatial distributions of street robbery. Both traditional and spatial statistics are used to analyse the results of the experiments.

The spatio-temporal dimension of the agents' routine activities is made explicit in the model by assigning them different types of activity spaces and movement. In the 'simple' version, agents are assigned a time to spend at home each day but once they leave home all movement is random. Agents in the 'temporal' version move randomly but have an assigned schedule that requires specific times be spent at work and other activities as well as at home. The 'activity space' version incorporates both temporal and spatial schedule constraints. Agents follow the same time schedule as in the temporal version but are also assigned specific locations that represent their home, workplace and activities. All movement is along the shortest path among these locations.

In the activity space version of the model the agent's activity spaces are based on the characteristics of Seattle. Their homes are assigned based on the distribution of population, and workplaces on the distribution of jobs. Activity locations are allotted based on the supply of services, retail and recreation. In this way the aggregate activity patterns of agents are representative of Seattle's population.

Fifteen experiments are run, each model version is run under five different experimental conditions. The experimental conditions essentially create five different artificial societies in which the only difference is the average amount of time that society spends away from home (e.g. under the first condition society spends 30% of their time away from home, in the second condition it spends 40% and so on). The individual agents spend differing proportions of time away from home, 8%, 17%, 42%, etc. It is only the average across all agents that must match the experimental condition being tested. This strategy provides a mechanism for systematically testing the effect of time spent away from home on street robbery rates.

Visualisation is critical to understanding how agent behaviour is translating into observed robbery patterns. Visualisation can occur either during or after the model run. Outcome measures can be displayed in the form of tables, graphs and/or maps (Figure 16.2).

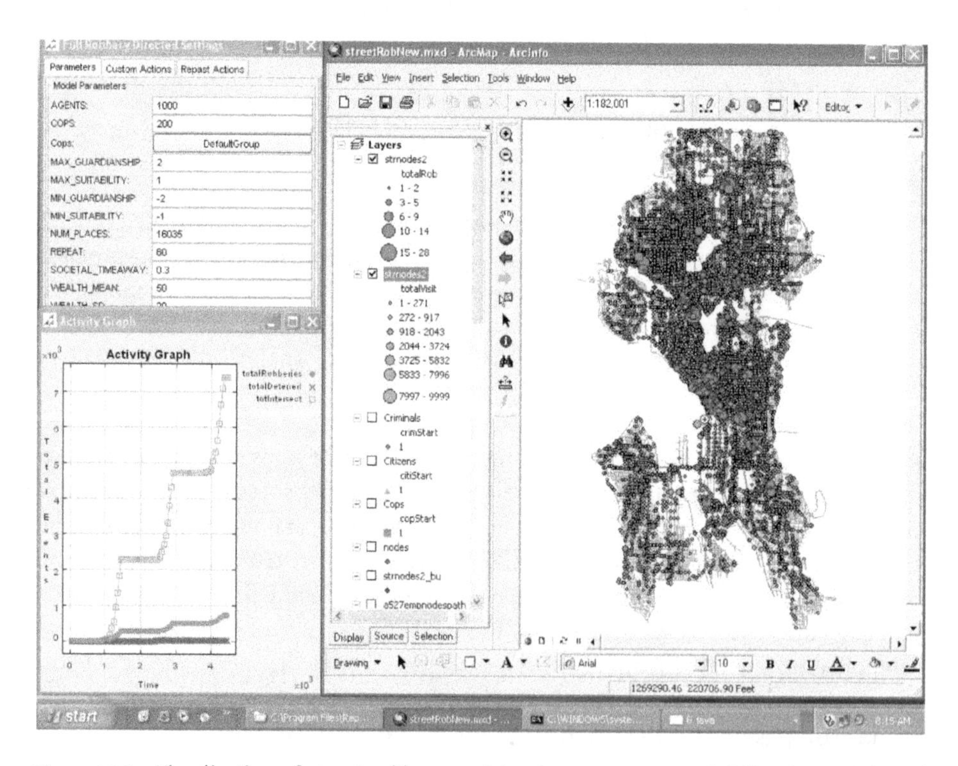

Figure 16.2 Visualisation of street robbery model outcome measures. A full-colour version of this figure appears in the plate section of this book.

16.4 Findings and significance of the research – comparing a simulated environment to the principles of routine activity theory

The results of this experiment on routine activity theory provide clear evidence on the veracity of the theory and on how the spatio-temporal nature of human activity influences the incidence and distribution of street robbery events. The results of the experiments provide support for routine activity theory's main premise in two of the three model versions. The model runs confirm that if the number of offenders and their motivation is held constant, as people spend more time away from home the number of street robberies will increase. The differences among the experimental conditions are significant in both the simple and temporal versions, but not for the activity space version.

In addition, the spatial patterns of the results vary significantly between versions and that variation remains as people spend more time away from home. Thus the results from systematically varying the spatio-temporal composition of routine activities show that both temporal and spatial constraints on routine activities play significant roles in determining the incidence of street robbery and should be included in future empirical studies that aim to tease out the role of routine human behaviour in robbery events. In particular, the findings demonstrate the importance of the street network and the distribution of opportunities in structuring travel and street robbery.

Methodological advances include the following:

- The research provides a demonstration of how simulation models offer a unique opportunity to formalise theories and then compare the theorised outcomes to the model outcomes.
- The research demonstrates the value of 'situating' simulation. Activity spaces are created based on the distribution of employment, housing, recreation, retail and services in the Seattle, Washington; thus, effectively incorporating the influence of land use on the activities of agents. Agents move along the vector GIS streets of Seattle. These advancements make it possible to compare the model outcome under three different assumptions of spatio-temporal activity constraints: randomness; random movement with temporal constraints; and both spatial and temporal constraints.
- The computational laboratory framework enables the first test of routine activity theory based on individual-level data. Although conducted using representational agents rather than empirical data, the computational laboratory framework permits a high level of scientific rigour to be applied to the design and testing of the model and the analyses of results.

In conclusion, the method used in this research represents an interim testing ground between the verbal formulation of the theory and the testing of theory with empirical data. Although these tests do not result in a determination of whether a theory is empirically valid, they do provide a way to strengthen the theory prior to empirical testing.

16.5 Comparing a simulated crime environment to reality

Although the goal of this model was not to accurately predict street robbery events, a comparison to the empirical spatial pattern of street robberies revealed that they shared common characteristics (Liu *et al.*, 2005). Both showed patterns of clustering (i.e. there were hotspots of robberies) and regardless of version, hotspots of crime were found in the central business district. These straightforward comparisons enhance the plausibility of the model's findings. Models that are developed to predict crime patterns, however, must undergo a more rigorous validation process that is still being debated. One challenge to model validation is that more than one model can produce the same results (O'Sullivan, 2004). How do we know which model best represents the underlying mechanisms? Another challenge specific to crime models is how to validate them when no reliable comparison data exist. Official crime data only represent a subset of crimes that occur (i.e. only crimes that are reported to the police and for which a crime report is filed). The question of how to validate simulation models is still an open one.

16.6 Implications for practice

Although the use of simulation modelling is still very much in its infancy within the environmental criminology research community, the potential implications for informing both policy and practice are numerous. For example, agent-based modelling offers a forum for exploring the impact of policy decisions. This is particularly important in the area of criminal justice because policy makers and practitioners must choose between competing strategies and often with limited evidence to guide their decisions.

In the area of crime prevention, policies suggested by Crime Prevention Through Environmental Design (CPTED) and situational crime prevention can be studied intensively. The physical design and access control strategies suggested by these approaches are sometimes very expensive to implement (e.g. closing streets, changing building design, etc.). Simulation saves money by identifying the strategies with the highest potential for success before investing in empirical testing. In addition, the quantitative output of the simulations can be used in preliminary cost-benefit analyses to evaluate new methods.

Opportunity theories postulate about the elements necessary for a crime to occur as well as the factors that bring those elements together at the same place and time. In a simulation, each one of those factors can be isolated and systematically manipulated while holding all the others constant. In this way confounding effects of variables can be separated. For example, the notion of guardianship is particularly important in opportunity theories and questions concerning the role of place managers as guardians could be tested. Place managers could be added into the model and the model run to examine how they changed the interactions and the patterns of crime that emerged. These types of experiments are essential to understanding the underlying mechanisms of crime patterns.

Agent-based models can also be used to test the effectiveness of different patrol strategies on crime prevention. For example, targeting police patrols to crime hotspots can be compared with a more general patrolling approach within police beats to test which is more effective at reducing crime. The optimum ratio of police officers to civilians for a particular jurisdiction can be explored by comparing the crime levels at varying force sizes. Displacement of crime and diffusion of benefits are other topics related to police practice that can be examined in new depth using simulation models.

The scope of ABM for theoretical testing and evaluation of strategies is very broad. This research provides a foundation on which more richly specified models can be developed. More advanced models have the potential to produce concrete, policy relevant findings to address the situational elements of crime. These elements can be changed more quickly and easily than ones involving the social and economic causes of crime and thus are more likely to provide immediate results.

16.7 References

Albrecht, J. (2005) A new age for geosimulation. *Transactions in GIS* **9**(4): 451–454.

An, L., Linderman, M., Qi, J., Shortridge, A. and Liu, J. (2005) Exploring complexity in a human–environment system: an agent-based spatial model for multidisciplinary and multiscale integration. *Annals of the Association of American Geographers* **95**(1): 54–79.

Bonabeau, E. (2002) Agent-based modeling: methods and techniques for simulating human systems. Paper presented at the *Arthur M. Sackler Colloquium of the National Academy of Sciences*, Irvine, CA.

Brantingham, P. L. and Brantingham, P. J. (2004) Computer simulation as a tool for environmental criminologists. *Security Journal* **17**(1): 21–30.

Brantingham, P. L. and Groff, E. R. (2004) The future of agent-based simulation in environmental criminology. Paper presented at the *American Society of Criminology*, Nashville, TN.

Brown, D. G., Riolo, R., Robinson, D. T., North, M. and Rand, W. (2005) Spatial process and data models: toward integration of agent-based models and GIS. *Journal of Geographic Systems* **7**: 25–47.

Cohen, L. E. and Felson, M. (1979) Social change and crime rate trends: a routine activity approach. *American Sociological Review* **44**: 588–608.

Dibble, C. (2001) *Theory in a complex world: GeoGraph computational laboratories*. Unpublished PhD dissertation, University of California Santa Barbara, Santa Barbara.

Eck, J. E. and Liu, L. (2004) Routine activity theory in a RA/CA crime simulation. Paper presented at the *American Society of Criminology*, Nashville, TN.

Epstein, J. M. and Axtell, R. (1996) *Growing Artificial Societies*. Washington DC: Brookings Institution Press.

ESRI (2005) *ArcGIS 9.1*. Redlands, CA: Environmental Systems Research Institute.

Gilbert, N. and Terna, P. (1999) *How to Build and Use Agent-based Models in Social Science*. Online Discussion Paper, http://web.econ.unito.it/terna/deposito/gil_ter.pdf [Retrieved 30 September 2003].

Gilbert, N. and Troitzsch, K. G. (1999) *Simulation for the Social Scientist*. Buckingham: Open University Press.

Groff, E. R. (2007) Simulation for theory testing and experimentation: an example using routine activity theory and street robbery. *Journal of Quantitative Criminology* **23**(2): 75–103.

Groff, E. R. (2008) Adding the temporal and spatial aspects of routine activities: a further test of routine activity theory. *Security Journal*.

Groff, E. R. (2007) 'Situating' simulation to model human spatio-temporal interactions: an example using crime events. *Transactions in GIS* **11**: 507–530.

Groff, E. R. (in press) Spatio-temporal aspects of routine activities and the distribution of street robbery. In *Artificial Crime Analysis Systems: Using Computer Simulations and Geographic Information Systems*, Liu, L. and Eck, J. (Eds). Hershey, PA: Idea Group.

Huisman, O. and Forer, P. (1998) Computational agents and urban life spaces: a preliminary realization of the time–geography of student lifestyles. Paper presented at *GeoComputation 98*, Bristol.

Liu, L., Wang, X., Eck, J. and Liang, J. (2005) Simulating crime events and crime patterns in RA/CA model. In *Geographic Information Systems and Crime Analysis*, Wang, F. (Ed.), pp. 197–213. Singapore: Idea Group.

Macy, M. W. and Willer, R. (2002) From factors to actors: computational sociology and agent-based modeling. *Annual Review of Sociology* **28**: 143–166.

North, M. J., Collier, N. T. and Vos, J. R. (2006) Experiences creating three implementations of the repast agent modeling toolkit. *ACM Transactions on Modeling and Computer Simulation* **16**(1): 1–25.

O'Sullivan, D. (2004) Complexity science and human geography. *Transactions of the Institute of British Geographers* **29**: 282–295.

O'Sullivan, D. and Haklay, M. (2000) Agent-based models and individualism: is the world agent-based? *Environment and Planning A* **32**(8): 1409–1425.

17 A crime mapping technique for assessing vulnerable targets for terrorism in local communities

Rachel Boba

17.1 Introduction

There are contributions that crime analysts in local police agencies can make to assess the vulnerabilities of their local communities based on the techniques and tools they currently use (e.g. see NCPE, 2006). Police agencies and crime analysts particularly focus on how and why crime opportunities exist in particular situations in order to seek solutions for crime prevention. The same can be done for potential terrorist activities. Ron Clarke and Graeme Newman (2006), scholars of situational crime prevention, in *Outsmarting the Terrorists*, provide concrete ways in which local law enforcement can focus crime analysis and crime mapping efforts towards understanding opportunities for terrorism, risk assessment and prevention. This work utilises criteria developed by Clarke and Newman (2006) for evaluating vulnerable targets, along with current crime mapping capabilities, to present a strategic technique for assessing areas within local communities based on target vulnerability for terrorism. A simulated case study is presented to illustrate the technique with the goal of assisting police practitioners and analysts with local-level counterterrorism efforts.

17.2 Assessing target vulnerabilities: two components

Clarke and Newman (2006) discuss four pillars of terrorist opportunity, which include targets, tools, weapons and facilitating conditions. They assert that these

Crime Mapping Case Studies: Practice and Research Edited by Spencer Chainey and Lisa Tompson

aspects of terrorist opportunities need to be understood in order to understand risk and develop prevention programmes. Specifically, the authors assert that local police 'must identify vulnerable targets, prioritise them for protection, analyse their specific weaknesses and provide them with protection appropriate to their risks' (Clarke and Newman, 2006, p. 4). Crime analysis and crime mapping can assist in all four aspects. However, this article focuses on one of these four aspects – assessing target vulnerabilities – and provides a specific technique that can be conducted utilising a geographical information system (GIS) and data available to most police agencies.

Generally, this technique uses a GIS, geographical data, local police experience and Clarke and Newman's (2006) criteria for vulnerable targets in order to develop a graduated-area map of a particular community that indicates high and low areas of risk for terrorism. Importantly, this technique can be conducted by a local-level crime analyst with the geographical data and software they currently have (Boba, 2005). The purpose of this technique is not to predict which individual target is the most vulnerable or when a particular target will be attacked, but to use crime mapping to present analysis results of risk by area based on multiple targets within an entire community, making it strategic in focus, not tactical.

Clarke and Newman (2006) emphasise that analysts and local police must understand why particular targets are more attractive to terrorists than others and insert themselves into the terrorists' decision-making process. In order to facilitate this, they have developed a systematic way of evaluating potential terrorist targets in any community which focuses on a target's attractiveness to terrorists. The authors developed the criteria to fit the acronym 'EVIL DONE', as an easy way to remember the concepts: Exposed, Vital, Iconic, Legitimate, Destructible, Occupied, Near and Easy.

These criteria are intended to be general in order to apply to a variety of communities, whether a large metropolitan area or a small rural community. For example, Mount Rushmore in South Dakota, USA, as a potential terrorist target, might be considered exposed, iconic, legitimate and destructible, but is not vital, occupied, near or easy because of its remote location. On the other hand, the World Trade Center Twin Towers in New York City contained all of these characteristics, which may be why they were targeted in several terrorist attempts over the years. Some geographical features that might be considered vulnerable include (Ronczkowski, 2004): nuclear power plants, ammonium nitrate repositories, airports, railroad tracks, mass transit lines, amusement parks, shopping malls, landmarks, research laboratories, dams, petroleum refineries, ports, government buildings, motorways, rivers, residential areas with high population density and major utility lines.

There are two components to this technique. The first component is the selection of an area at risk, based on the location of a particular target or type of target. Instead of looking at each target individually, as is done for tactical analysis and response, a jurisdiction is broken down into small predetermined areas so target vulnerabilities can be analysed together, areas prioritised and strategic prevention efforts focused in particular areas of a community. The GIS software allows the analyst to select predetermined areas through several different methods. They include (specific examples of these will be provided in the practical example below):

- *Intersection*: this allows the analyst to select an area based on whether it intersects at any point with a particular target or type of target.
- *Within a distance*: this allows the analyst to select an area based on its proximity to a particular target or type of target.
- *Completely within*: this allows the analyst to select an area based on whether the target is completely within the boundaries of the area.
- *Contain any part*: this allows the analyst to select an area based on whether any part of the target is within the area, which is useful when a target is larger in area (e.g. nuclear power plant).

The second component of the technique is assigning a score to a selected area based on a particular target's or type of targets level of vulnerability. The EVIL DONE schema is used to score selected areas based on the nature of targets within that area. Each of the eight components can be assigned a score of 1 so each target would be scored according to how many of the criteria it has on a scale of 0 to 8. Taking the Mount Rushmore example again, it had four of the eight criteria thus any areas selected based on this target would be assigned a four. The score of a particular target or type of target would require consideration by those in the agency with expertise in terrorism and target vulnerability.

The analysis could also be more complex by assigning a weight to each component of the EVIL DONE schema, instead of considering them equal, based on importance. For example, if a target is considered 'vital' it may be given a value of 2 whereas 'easy' might remain as a 1 since many targets are easy and this criterion is not as important. This too would be decided by those in the agency with expertise in this area.

To combine these two components of the technique, the types of targets, represented as layers in a GIS, would be scored separately and summed for each area. The cumulative score for each area would be thematically shaded gradually based on that score relative to the other areas. The thematic classification (e.g. natural breaks, standard deviation, equal interval) used for graduation would depend on the purpose of the analysis and nature of the data and should be selected by the analyst (see Boba, 2005). The result would be a thematic shading of areas with the darkest (highest score) implying a higher risk. This technique and its components are illustrated in the following simulated case study.

17.3 Assessing target vulnerabilities: a hypothetical case study

Figure 17.1 depicts a map of an imaginary city with several types of geographical features that might be potential terrorist targets. For the sake of simplicity, only major research facilities (e.g. biological research), rivers, schools, railroads, government buildings, streets and zones (i.e. predetermined generic areas which in practice may be based on police beats, census block groups or census tracts, for example) are included in the case study. Particular types of targets and the specific geographical area unit of analysis would depend on the individual community assessment and available data.

What follows is a series of maps that illustrate the different ways the predetermined areas can be selected within a GIS and assigned scores according to the EVIL DONE criteria. Note that each component of the EVIL DONE criteria is valued as 1 and the score is estimated using a general assessment of that type of target. In practice, the assessment would be made based on the type of location generally but also the specific nature of that particular target in a particular community.

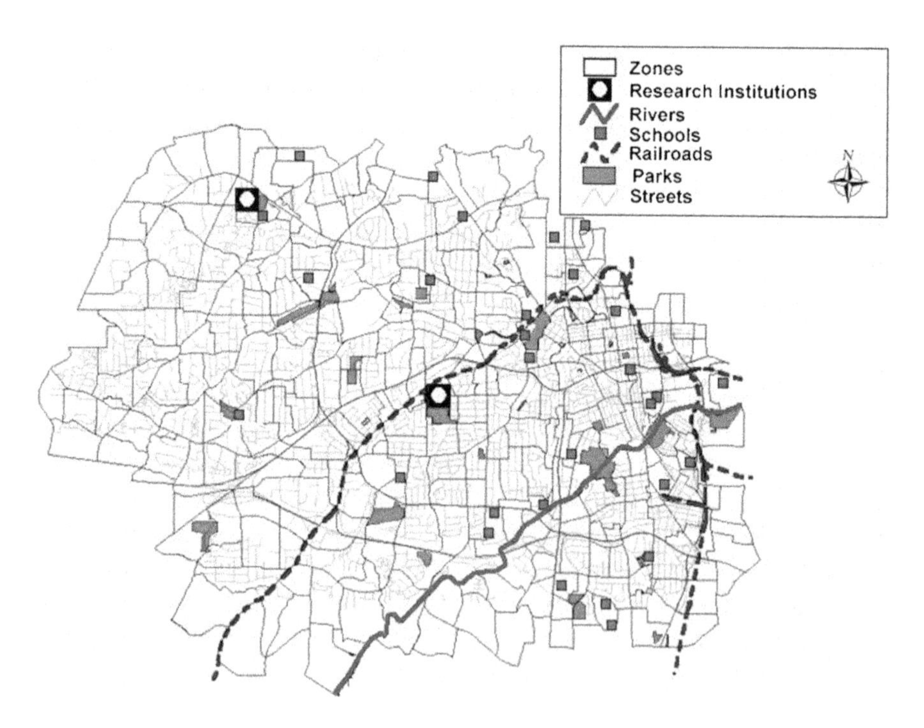

Figure 17.1 Map features of hypothetical city.

Figure 17.2a shows the river feature with the selected zones through which it flows, or 'intersects' it. The score for these zones is 4, based on the fact that rivers are generally exposed, legitimate, near (in this case), and easy, but not necessarily vital, iconic, destructible or occupied. Obviously, the score of a river might be higher or lower in another situation. Another consideration is that individual areas along a river might be scored differently – for example, based on whether a river segment has significant housing along its banks.

Figure 17.2b illustrates a railroad feature with selected zones with borders that are '500 feet' from the railroad. The selection method is different than that of the river because it is selecting a zone within a distance of the feature, not just the zones that intersect the feature. This might be done for a railroad because these areas would provide access to the railroad or might be populated. The score for a railroad generally is 6 because it is exposed, vital, legitimate, destructible, near, and easy, but may not be iconic or occupied. Like the river example, individual areas along the railroad might also be scored differently based on specific characteristics of that area.

Figure 17.2c shows the locations of schools (as points) and the zones shaded are only those in which a school is 'completely within' (the centre of the school symbol indicates which zone will be shaded). The score for these schools is 8, as schools generally have all the EVIL DONE characteristics. The analysis within a specific community may vary the score by characteristics and vulnerability of each school.

A further way to select zones around a feature is illustrated in Figure 17.2d, which depicts parks, not as points, like schools in the previous map, but as polygons (representing the expanse of the park) so that the entire area of the park is considered. Using areas instead of points might also be used for school locations as often the school campuses can cover a large area that point features fail to spatially represent. Comparing Figure 17.2c with Figure 17.2d, one can see the difference in the number of zones shaded grey. These zones are shaded if they 'contain any part' of the park polygon. The score assigned to these areas containing any part of a park is 4 because parks are generally exposed, legitimate, near, and easy. Scores may, however, vary by season or specific dates when a park may become iconic and occupied, for example in the USA during a Fourth of July (Independence Day) fireworks celebration.

A final example, illustrates zones 'within ½ mile' of a research institution that are selected in Figure 17.2e. This selection criterion might be used for a location that if targeted by terrorists would affect areas a distance from it, because of chemical fallout or other reasons. The score of the selected areas is 6 because they are exposed, legitimate, destructible, occupied, near and easy.

Once the zones have been selected and scores assigned for each geographical feature, the scores for each zone are added together within the GIS and depicted

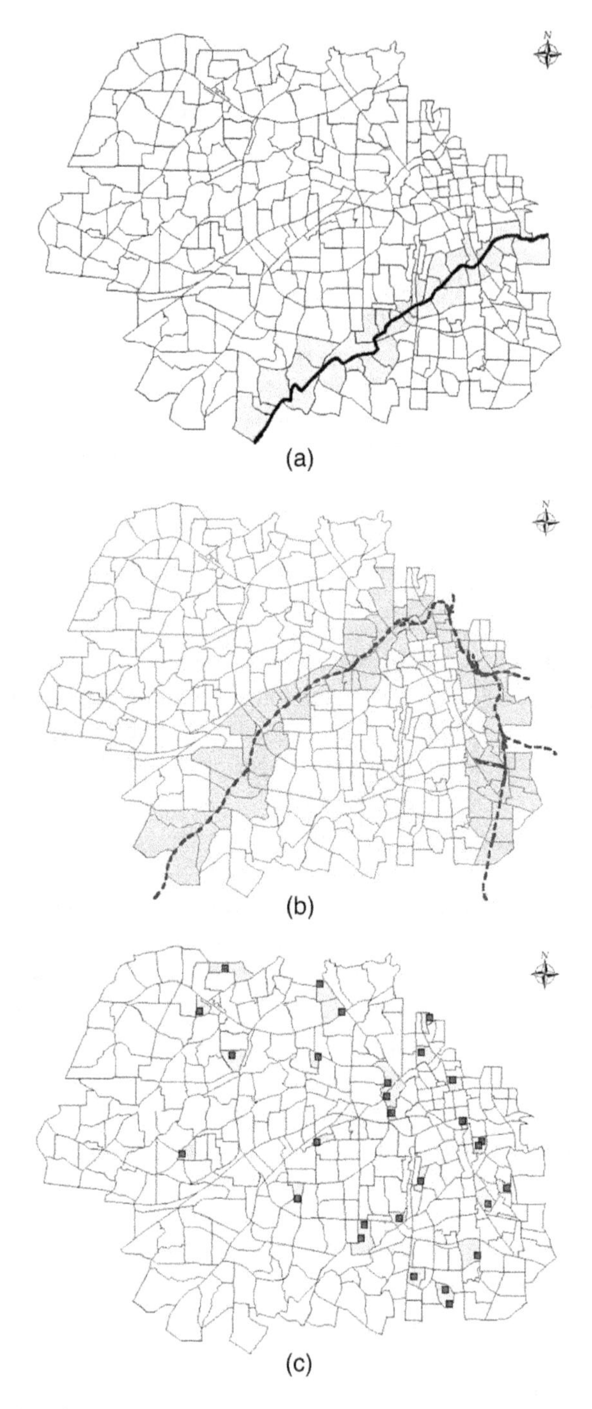

Figure 17.2 (a) River feature, (b) railroad feature, (c) school feature, (d) park feature, and (e) research institution feature scored and mapped against the EVIL DONE (Exposed, Vital, Iconic, Legitimate, Destructible, Occupied, Near and Easy) criteria.

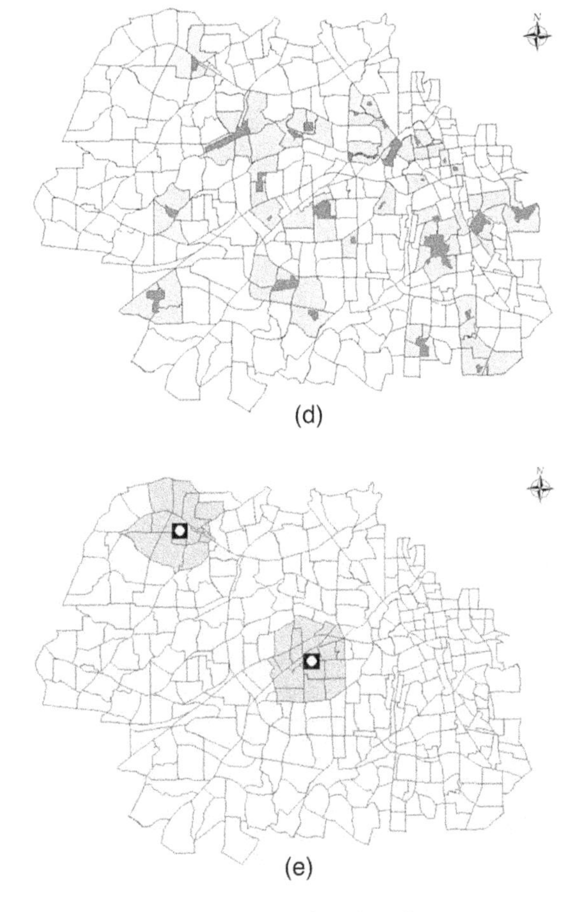

Figure 17.2 (*Continued*)

on a map to present target vulnerability within the entire community. Figure 17.3 illustrates the summary scores of the previous analysis. The map can now be used to prioritise strategic prevention efforts and resource allocation within a community.

To reiterate, these maps illustrate a geographical method that can be conducted by a local crime analyst to assist with analysis of target vulnerability for potential terrorist events. It could help a police agency as well as the city government locate response efforts as well as prioritise target hardening. The size and actual section of the predetermined geographical units would be based on knowledge of the specific jurisdiction, its topography and proximity to other features. For example, in a jurisdiction with less square mileage, the geographical units might be relatively small. Or, the analyst may use the GIS to place an arbitrary grid over the map to

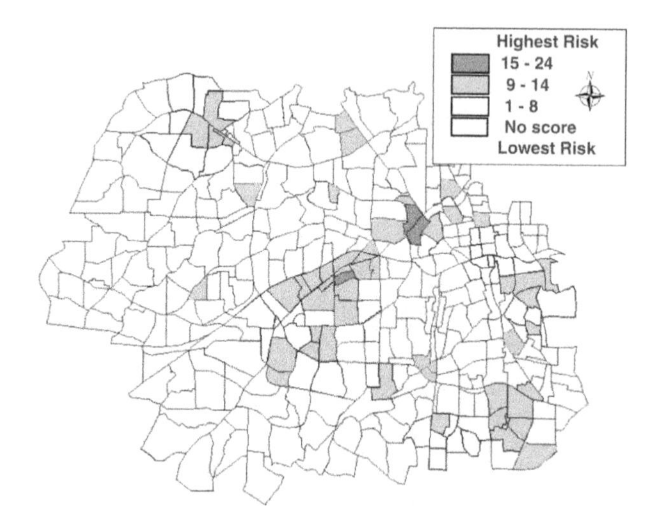

Figure 17.3 Risk analysis results map based on the EVIL DONE (Exposed, Vital, Iconic, Legitimate, Destructible, Occupied, Near and Easy) criteria.

signify areas to allow a comparison of areas of the same size. The scores based on the EVIL DONE schema would also depend on local situation and opportunities and could even vary within feature type. For example, one research facility working with anthrax might have a higher score than one that does not.

Lastly, this method should not be viewed as static in that one map created at one particular point in time should not be considered as sufficient. As was noted above, events throughout the year, such as a Fourth of July celebration, may change the analysis for that month. Targets and their vulnerabilities are continuously changing making it necessary to continually update the analysis. However, the ease of this technique and the use of existing software, data and analysts' skills make a fairly straightforward task to repeat.

17.4 Considerations

There are a few words of caution when using this technique. The first is that it has not been empirically tested, so decision making should be based on these results along with other factors. The second is that many of the decisions that need to be made within the analysis are subjective and should be made by a team of personnel with expertise in spatial analysis and counterterrorism. A team approach is essential for an effective product. Lastly, this method does not put an 'X' on the spot where terrorism is likely to happen, but provides a way for a police department to prioritise target hardening as well as education of citizens and business owners in the most vulnerable areas of their communities. This method provides local police agencies with a basic crime mapping and analysis technique that can make a reasonable attempt at assessing vulnerability of targets for terrorism in their community.

17.5 References

Boba, R. (2005) *Crime Analysis and Crime Mapping*. Thousand Oaks, CA: Sage Publications, Inc.

Clarke, R. V. and Newman, G. (2006) *Outsmarting the Terrorists*. Portsmouth, NH: Greenwood Publishing Group.

NCPE (2006) Briefing paper on *Neighbourhood Policing and the NIM* (online). www .neighbourhood policing.org.uk

Ronczkowski, R. (2004) *Terrorism and Organized Hate Crime: Intelligence Gathering, Analysis, and Investigations*. Boca Raton, FL: CRC press.

18 Interactive Offender Profiling System (IOPS)

David Canter and Donna Youngs

18.1 Introduction

The Interactive Offender Profiling System (IOPS) has its origins in the International Centre for Investigative Psychology (ICIP) studies of the behavioural and spatial patterns that can be identified within offenders' actions. Developed to support the next generation of software tools for police and law enforcement analysts, IOPS integrates large police databases at speed, drawing directly on research findings to:

- link crimes;
- prioritise suspects;
- build catalogues of offenders' geobehavioural profiles;
- generate potential TICs (that is, further offences that are 'taken into consideration');
- explore co-offending networks;
- indicate locations for intelligence gathering;
- allow hotspot analysis;
- provide geocoding capabilities;
- provide powerful mapping capabilities from its ArcGIS (or similar) platform.

Crime Mapping Case Studies: Practice and Research Edited by Spencer Chainey and Lisa Tompson

The original vision of the IOPS system (Canter, 2004) emerged out of twenty years of research. The following illustrative operational scenario indicates one possible utilisation of that system.

- A member of the public reports a burglary. The address and crime details are recorded from a scene of crime visit or other remote information source. This is integrated into existing databases and related to online detailed maps (and/or aerial photography) and made available to dedicated crime analysts in a local intelligence unit or specialist intelligence section.
- The digitised geographical information is supported by crime data and a set of analysis functions that allows rapid comparative analysis, facilitated by interaction with the background. This enables the crime analyst to construct inferences concerning possible linked burglaries in the area and possible suspects. This leads to the proposal of a number of possible offenders and their likely residential locations. Information on known recipients of stolen goods and illegal drugs suppliers is also available.
- By interrogating the system the analyst alerts a local police patrol to observe an area for the criminal and to pay particular attention to three most likely suspects with a note of where they are currently living.

18.2 An integrated operational system

The first version (IOPS 1.0) was developed for the UK's Metropolitan Police using software engineers at the Kelvin Institute of Strathclyde University. It draws on large police datasets and uses this to perform analysis using the geographical profiling software, Dragnet, to explore offending spatial behaviour and generate geographical profiles that are then integrated with multidimensional scaling-based visual models of offender *modus operandi* (MOs). These and other analyses are combined in different ways to directly address the operational questions analysts and detectives face in investigations. The system has the potential to assist many investigations of serial, violent and volume crime by:

- helping to determine what aspects of the crime should be highlighted to guide the investigation;
- providing the basis for generating inferences about the distinguishing characteristics of the offender;
- assisting in the determination of which crimes are likely to be linked;
- offering possibilities of which property crimes may be linked to each other or to violent crimes;

- indicating geographical locations in which further intelligence may be forthcoming;
- proposing geographical locations of particular significance to an offender;
- generating priority rankings for suspects and other nominals using both geographical matching and *modus operandi* matching – the system searches for known offenders and puts them in rank order on the basis of their location and also their '*modus operandi*'.

The IOPS system has a number of key components that interact with each other. These are explained in the next sections.

18.2.1 Direct mapping of offences and offenders

ArcGIS is used to represent on a map all offences within any given offence type and period along with known relevant offenders within the police databases. The detective or analyst may interact with this map to select a subset of these offences according to various operational criteria. The actions committed within these crimes are drawn out from the police information system. These actions are then subject to subsequent analysis to create *modus operandi* (MO) Heat Maps.

18.2.2 Modus operandi (MO) Heat Maps

Modus operandi Heat Maps are a development of Investigative Psychological studies identifying the key variations in offending style within crimes from burglary to serial rape (for a review of this work see Canter and Youngs; 2003, 2008). The MO Heat Maps consist of Smallest Space Analyses of offence activity in which the frequencies of each action are indicated by a colour code. This gives rise to regions of similar frequency having similar colours and therefore reveals the structure of the overall pattern of behaviours. Figure 18.1 illustrates this.

18.2.3 Comparative case analysis

On the basis of the interactive selection of actions from the MO Heat Maps, it is possible to compare a subset of offences selected from the original set of offences and represent these as points in a space such that the closer the points representing the offence, the more similar the offences in terms of the originally selected actions (see Figure 18.2). This configuration is open to interactive use such that crimes that are similar can be identified on the screen and then portrayed on a map for further geographical analysis.

Figure 18.1 A *modus operandi* Heat Map. The points in this illustration are criminal actions. The closer together the actions, the more likely they are to co-occur in any crime. A full-colour version of this figure appears in the plate section of this book.

18.2.4 Prioritising nominals

A further operationally informative exploration that IOPS allows draws initially on a geographical profiling analysis of the crimes to generating a prioritised map of the likely base locations of possible offenders (Canter and Youngs, 2008). This is then superimposed on the original map of known offenders, such that all those nominals residing within the broad area indicated by the geographical profiling analysis can be identified. The IOPS system produces a map with possible suspects indicated by the crosses (see Figure 18.3).

Figure 18.2 A comparative case analysis, drawn from a selection of offences from the *modus operandi* Heat Map, shown as each offence with a police record number that they could have been allocated within a given police database.

18.2.5 Geographical and modus operandus prioritisation

For any unsolved offence, IOPS then examines all known offenders in the system in terms of the closeness of match with the MO of the offence and the geographical profile. This is used to produce a suspect prioritisation table.

18.2.6 Social network analysis

A further capability of the system is the representation of known associations between offenders as links directly on a map, allowing additional analysis of other potentially linked crimes (see Figure 18.4)

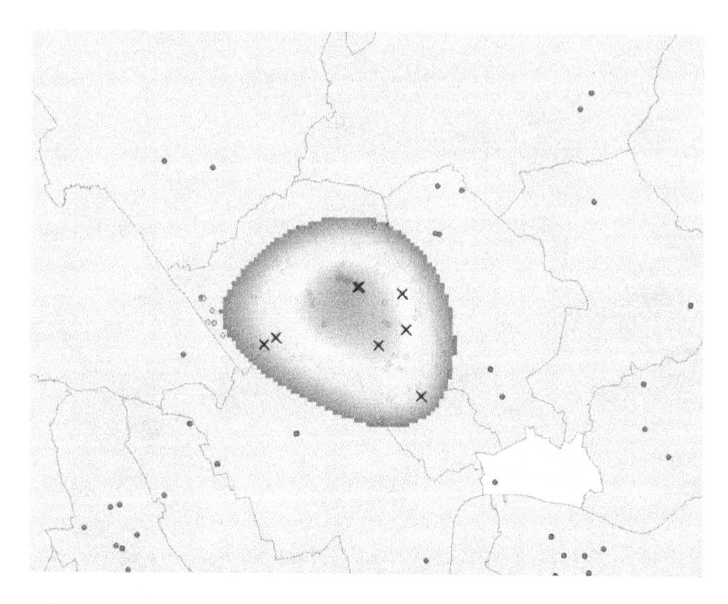

Figure 18.3 A geographical profile of a potential offending series (the cross symbols), shown with possible suspects (the light dots would represent offenders living in a specified jurisdiction, the dark dot symbols would represent offenders living outside a specified jurisdiction). A full-colour version of this figure appears in the plate section of this book.

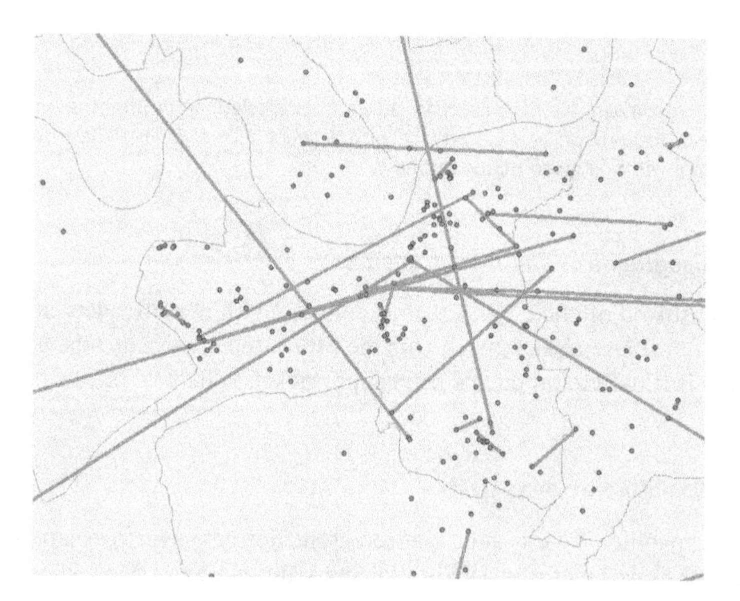

Figure 18.4 This map shows linkages of 'known' associations between potential offenders, helping to support additional analyses of their potentially linked crimes.

18.3 The potential of the interactive offender profiling system

Clear up rates for volume crime offences are remarkably low (e.g. ICIP studies indicate average clear up rates of 9.3% for street robbery and 9.7% for residential burglary). Significant improvements in the effectiveness of the police in crime detection (as well as reduction) are now potentially possible by exploiting recent advances in offender profiling and geographical profiling (Canter and Youngs, 2008) to develop crime investigation software that allows crime analysts and investigators in a systematic fashion, in real time and utilising the existing police databases, to:

- produce a summary of characteristics of offenders who have committed similar crimes (an 'Offender Profile') on the basis of the MO;
- identify the likely area of residence of a perpetrator based on the location of the offences, as well as the land use, environmental features and the aggregate crime patterns in an area (Advanced Geographical Profiling);
- integrate the Offender Profile with the Geographical Profile to prioritise suspects within a subset of known offenders held within the police database (Geobehavioural Profiling);
- link further unsolved offences to a common offender (Comparative Case Analysis);
- identify likely co-offenders and map out 'criminal networks' (Geographical Social Network Analysis);
- use time and sequence analysis of MO and geographical development to predict future activity;
- from a crime reduction perspective, continually map ongoing changes in the geographical patterning of MOs, such that responsive, area-specific initiatives are possible to counter particular MOs and offender patterns.

The analysis of the geobehavioural and social aspects of crimes in a real-time police environment also provides a powerful research context. The system therefore has the capability of continually learning in a research mode about criminal activity and feeding this back into operational use. The IOPS system thus grows out of, and feeds back into, the research environment of the International Centre for Investigative Psychology within which it has been developed.

18.4 Acknowledgement

We are grateful to Freya Newman from the International Centre for Investigative Psychology for her help in preparing the material for this paper.

18.5 References

Canter, D. (2004). *The System: Towards an Interactive Offending Profiling System*. Internal document, International Centre for Investigative Psychology.

Canter, D. and Youngs, D. (2003). Beyond offender profiling: the need for an investigative psychology. In *Handbook of Psychology in Legal Contexts*, Bull, R. and Carson, D. (Eds), pp. 171–206. Chichester: Wiley.

Canter, D. and Youngs, D. (2008). *Principles of Geographical Offender Profiling*. Aldershot: Ashgate.

Canter, D. and Youngs, D. (2008). *Applications of Geographical Offender Profiling*. Aldershot: Ashgate.

Canter, D. and Youngs, D. (2008). *Investigative Psychology*. Chichester: Wiley.

Index

Crime Mapping Case Studies: Practice and Research Edited by Spencer Chainey and Lisa Tompson

www.ingramcontent.com/pod-product-compliance
Lightning Source LLC
Chambersburg PA
CBHW082045280125
20788CB00044B/50